T0135541

Coupled Electromagnetic Field/Circuit Simulation: Modeling and Numerical Analysis

Inaugural-Dissertation

zur

Erlangung des Doktorgrades

der Mathematisch-Naturwissenschaftlichen Fakultät

der Universität zu Köln

vorgelegt von

Sascha Baumanns

aus Kamp-Lintfort

Köln 2012

Berichterstatter/in: Prof. Dr. Caren Tischendorf
                       Prof. Dr. Ulrich Trottenberg

Tag der mündlichen Prüfung: 19.06.2012

Bibliografische Information der Deutschen Nationalbibliothek

Die Deutsche Nationalbibliothek verzeichnet diese Publikation in der
Deutschen Nationalbibliografie; detaillierte bibliografische Daten sind
im Internet über http://dnb.d-nb.de abrufbar.

ISBN 978-3-8325-3191-1

Logos Verlag Berlin GmbH
Comeniushof, Gubener Str. 47,
10243 Berlin
Tel.: +49 (0)30 42 85 10 90
Fax: +49 (0)30 42 85 10 92
INTERNET: http://www.logos-verlag.de

Kurzzusammenfassung

Heutige Schaltungsmodelle verlieren in der Schaltungssimulation aufgrund der rasanten technologischen Entwicklung, Miniaturisierung und höherer Komplexität von integrierten Schaltungen zunehmend ihre Gültigkeit. Dies motiviert die direkte Kombination von Schaltungssimulation mit Bauelementesimulation für kritische Schaltungsteile.

In dieser Arbeit betrachten wir ein Modell von partiellen Differentialgleichungen für elektromagnetische Bauelemente - modelliert durch die Maxwell-Gleichungen - gekoppelt mit differential-algebraischen Gleichungen, welche die einfachen Schaltungselemente einschließlich Memristoren und die Topologie der Schaltung beschreiben.

Wir untersuchen das gekoppelte System nach Diskretisierung der Maxwell-Gleichungen in einer Potentialformulierung im Ort durch die Finite Integration Technik, die eine gängige Methode in der Praxis ist. Das ortsdiskretisierte gekoppelte System ist als differential-algebraische Gleichung mit einem proper formulierten Hauptterm modelliert. Es werden topologische Bedingungen sowie Modellierungsbedingungen, die sicherstellen, dass der Index der differential-algebraischen Gleichung nicht größer als zwei ist, präsentiert. Es zeigt sich, dass der Index abhängig von der gewählten Eichbedingung für die Maxwell-Gleichungen ist.

Für die erfolgreiche numerische Integration von differential-algebraischen Gleichungen spielt die Index-Charakterisierung eine entscheidende Rolle. Der Index kann als Maß für die Empfindlichkeit der Gleichung gegenüber Störungen der Eingangsfunktionen und numerischer Schwierigkeiten, wie der Berechnung von konsistenten Anfangswerten für Zeitintegration, gesehen werden.

Wir verallgemeinern Indexreduktionstechniken für den Traktabilitätsindex für eine allgemeine Klasse von differential-algebraischen Gleichungen. Mit Hilfe der Indexreduktion erhalten wir lokale Lösbarkeits- und Störungsaussagen für differential-algebraische Gleichungen mit einem proper formulierten Hauptterm vom Index-2, und wir geben einen Algorithmus an, um konsistente Initialisierungen für das ortsdiskretisierte gekoppelte System zu bestimmen.

Schließlich werden die Ergebnisse durch numerische Experimente verifiziert.

Abstract

Today's most common circuit models increasingly tend to loose their validity in circuit simulation due to the rapid technological developments, miniaturization and higher complexity of integrated circuits. This has motivated the idea of combining circuit simulation directly with distributed device models to refine critical circuit parts.

In this thesis we consider a model, which couples partial differential equations for electromagnetic devices - modeled by Maxwell's equations -, to differential-algebraic equations, which describe basic circuit elements including memristors and the circuit's topology.

We analyze the coupled system after spatial discretization of Maxwell's equations in a potential formulation using the finite integration technique, which is often used in practice. The resulting system is formulated as a differential-algebraic equation with a properly stated leading term. We present the topological and modeling conditions that guarantee the tractability index of these differential-algebraic equations to be no greater than two. It shows that the tractability index depends on the chosen gauge condition for Maxwell's equations.

For successful numerical integration of differential-algebraic equations the index characterization plays a crucial role. The index can be seen as a measure of the equation's sensitivity to perturbations of the input functions and numerical difficulties such as the computation of consistent initial values for time integration.

We generalize index reduction techniques for a general class of differential-algebraic equations by using the tractability index concept. Utilizing the index reduction we deduce local solvability and perturbation results for differential-algebraic equations having tractability index-2 and we derive an algorithm to calculate consistent initializations for the spatial discretized coupled system.

Finally, we demonstrate our results by numerical experiments.

Acknowledgement

This thesis was written during my employment at the Chair of Mathematics/Numerics at the University of Cologne.

First of all special thanks are due to Prof. Dr. Caren Tischendorf for all the interesting tasks, her guidance, faith and fruitful discussions. It was a great opportunity to have been member of her working group.

I also would like to thank Prof. Dr. Ulrich Trottenberg for serving as referee of my thesis, Prof. Dr. Rainer Schrader, who agreed to be the chairperson of my doctoral committee, and Dr. Roman Wienands to be a committee member.

I owe gratitude to my colleagues at the University of Cologne, former and present, - in particular Lennart Jansen, Michael Matthes, Dr. Michael Menrath, Prof. Dr. Monica Selva Soto and Leonid Torgovitski - for all the discussions on differential-algebraic equations and mathematics in general, but also for being friends with whom to work with has always been a pleasure.

Also I would like to show my gratitude to the colleagues of the Cologne branch of the Fraunhofer-Institute for Algorithms and Scientific Computing SCAI for providing a pleasant work environment, in particular Dr. Tanja Clees, Dr. Bernhard Klaassen and Clemens-August Thole.

My colleagues in Wuppertal and Darmstadt - Dr. Andreas Bartel, Prof. Dr. Markus Clemens and Prof. Dr. Sebastian Schöps - all played an important role in this research. I thank them all for the interesting and encouraging discussions on Maxwell's equations.

Dr. Wim Schonemaker from MAGWEL owes my thanks for sharing his thoughts and experience in electromagnetic device simulation with me.

I would also like to thank Yvonne Havertz for proofreading the thesis.

Special thanks are also directed to my parents who have always supported me. I am grateful to my family and the Glaser family for their outstanding support.
Finally, I wish to thank my fiancée Judith Glaser for all her motivational skills, her drive and her patience.

I am indebted to the ICESTARS (FP7/2008/ICT/214911) project, funded by the EU's Seventh Framework Programme for Research, and the SOFA project (03MS648A), funded by the German Federal Ministry of Education and Research (BMBF) programme "Mathematik für Innovationen in Industrie und Dienstleistungen".

Frechen, August 1, 2012   Sascha Baumanns

Contents

List of Figures

List of Tables

1 Introduction

Insofern sich die Sätze der Mathematik auf die
Wirklichkeit beziehen, sind sie nicht sicher, und
insofern sie sicher sind, beziehen sie sich nicht auf
die Wirklichkeit.

ALBERT EINSTEIN, 1879-1955

In various fields such as automotive industry or telecommunication technological progress
is mainly driven by a rapid development of integrated circuits. The enormous growth of
performance is based on a higher complexity and packing density of integrated circuits
as well as decreasing spatial scales and increasing frequencies of electronic devices.
The miniaturization of the circuits causes an increasing power density, which in turn
makes it necessary for the prediction of the circuits behavior to take, amongst others,
into account heating effects, electromagnetic fields and an accurate switching behavior
of semiconductors.

A common tool to predict the behavior of circuits and to reduce the costs of development
is circuit simulation. Due to the complexity, which arises from up to millions of circuit
elements it is absolutely necessary to keep the model sizes as low as possible. The
consequences are contradicting demands in circuit simulation: On the one hand the
physical behavior of the circuit needs to be described correctly whereas on the other
hand the computing time must be reasonably small.
A well established approach, which tries to fulfill both requirements is the modified nodal
analysis, see [CL75, CDK87, DK84]. The modified nodal analysis models the circuit with
basic elements only, such as capacitors, resistors, inductors, voltage and current sources.
Complex elements such as semiconductors or even conductors and their interactions,
respectively, are modeled by equivalent circuits consisting of basic elements only. The
modeling of equivalent circuits in an appropriate manner is a challenging task leading
to hundreds of model parameters, see [DF06].

Due to decreasing spatial scales and increasing frequencies the device behavior is also
influenced by the surrounding circuitry, for example, by inductive coupling. It happens
with ever greater frequency that these equivalent circuits are not accurate enough and
a refined modeling of a particular device is necessary. Consequently, for complex circuits it is recommended to directly combine circuit simulation with device simulation for
particular devices. However, due to up to millions of circuit elements belonging to one
circuit we are restricted to equivalent circuits for most devices. There is a wide range

of modeling levels from linear and nonlinear equations to partial differential equations depending on the effects to be described, for example, heating [Bar04, Cul09], semiconductor behavior [Tis04, Bod07] and electromagnetic fields [Gün01, Ben06, Sch11].

From the circuit designer point of view not only new manufacturing technologies have a great impact on future integrated circuits but also the development of new circuit elements. Such a new element that most likely will be of huge impact is the memristor. In 1971 Leon Chua postulated the theory of such an element to be existing, but only in 2008 the first physical model was released by HP Labs, see [Chu71, SSSW08]. Apart from the three basic elements, namely, the capacitor, the resistor and the inductor, already discovered in the 18th and 19th century, the memristor is considered the fourth basic element. This holds true as the behavior of the memristor cannot be reproduced by any circuit using only the other three basic elements, see [Chu71].

The circuit models including refined devices and memristors lead to systems composed of linear, nonlinear and ordinary differential equations after applying a method of lines. For a reliable simulation of these systems we are interested in the perturbation sensitivity. Thus, modeling these systems as differential-algebraic equations with a properly stated leading term is an appropriate approach since for certain classes of such differential-algebraic equations it has been shown that backward differentiation formulas and Runge-Kutta methods are stability preserving, see [HMT03a, HMT03b].

There are several different index concepts to characterize a differential-algebraic equation. All concepts are a measure of the difficulties to be found in the numerical simulation such as sensitivity to input perturbations. A direct measure of this sensitivity is the perturbation index, which takes perturbations of the right hand side into account. These perturbations result, for instance, from round-off and Newton method errors. However, the perturbation index in general is difficult to determine. For our investigations we choose the tractability index concept to determine the differential-algebraic equation's sensitivity with respect to perturbations.

This thesis is based on three basic issues, namely, differential-algebraic equation theory, structural investigations of circuits including memristors and structural investigations of circuits including refined electromagnetic devices modeled by Maxwell's equations.

The first basic issue is the differential-algebraic equation theory, which is the basis for our later analysis of the extended circuit models. We familiarize with differential-algebraic equations with a properly stated leading term and the index concept up to index-2 used in this thesis. For a complete overview on that topic we refer to [LMT13]. The differential-algebraic equation analysis is guided by a generalization of the index reduction technique by differentiation of differential-algebraic equations without a properly stated leading term to differential-algebraic equations with a properly stated leading term lowering the index down from two to one. We show that the index reduced differential-algebraic equation has index-1 and a properly stated leading term. Utilizing the index reduction we deduce local solvability and perturbation index-2 for differential-algebraic equations having index-2 from well-known index-1 results given in [LMT13]. One of the difficulties in solving differential-algebraic equations numerically is to determine a consistent initial value to start the integration. To solve this problem we derive an algorithm to calculate consistent initializations for differential-algebraic equations having index-2 and a properly stated leading term by an approach, which is a generalization of the results of [Est00]. All results are derived under sufficient structural conditions.

The second basic issue is the structural investigation of circuits including memristors. We introduce the characteristic equations and topology for capacitors, resistors, inductors, voltage and current sources known from literature, [CDK87, DK84], and, in addition, for the memristor, [Chu71]. For circuits without memristors we introduce the modified nodal analysis and well-known topological index results of [Tis99, ET00]. For a potential use of the memristor in circuit simulation we need to embed the memristor in actual circuit models. The nodal analysis method has already been extended by memristor models. The index of the resulting differential-algebraic equation is investigated in [Ria10]. In this thesis we extend the modified nodal analysis by memristor models and the structural properties of resulting differential-algebraic equation are presented, which leads to an extension of the topological index results for circuits without memristors.

The third basic issue is the structural investigation of circuits including refined devices modeled by Maxwell's equations. First, we investigate Maxwell's equations. The partial differential equations have been postulated by James Clerk Maxwell in the middle of the 19th century and form the basic of the modern theory of electromagnetics, see [Max64]. We take Maxwell's equations in a potential formulation into account which are very popular in a broad filed of applications. Various spatial discretizations have been studied, see [BP89, StM05, Cle05, CMSW11], and common spatial discretizations are the cell method [Ton01], a finite-volume method [MMS01] and variants of the finite-element method [Ned80, Bos98, Göd10]. In the work presented we opt for the finite integration technique introduced in 1977 by Thomas Weiland [Wei77] for spatial discretization. Weiland generalized a finite-difference time-domain-scheme of Kane Yee [Yee66], also known as leap-frog scheme, to solve Maxwell's equations. The potential approach results in a suitable description of Maxwell's equations and provide a natural link to the concept of potential differences used in circuit simulation. However, the potentials are not uniquely defined and a gauge condition is needed, see [Jac98, Bos01]. For the finite integration technique, grad-div formulations based on the Coulomb gauge are already well known, see [CW02, BCDS11]. In this thesis we introduce a new class of gauge conditions in terms of the finite integration technique motivated by a Lorenz gauge formulation. After spatial discretization we analyze the structural properties of resulting differential-algebraic equation formulated with a properly stated leading term. It turns out that the index of the differential-algebraic equation depends on the chosen gauge condition but without exceeding index-2. To concentrate the link to circuit simulation a suitable boundary excitation and current formulation is deduced. Next we investigate coupled electromagnetic device/circuit models with spatially resolved electromagnetic devices, where the electromagnetic devices are described by Maxwell's equations in a potential formulation spatially discretized by the finite integration technique. In literature coupled magnetoquasistatic device/circuit models are investigated in [HM76, KMST93, DHW04, DW04] and the index of the resulting differential-algebraic equations is discussed in [Tsu02, Ben06, BBS11, Sch11] using certain conductor models and different circuit configurations. These results extend the topological index conditions for the modified nodal analysis given in [Tis99, ET00]. Our index analysis for coupled electromagnetic device/circuit models is not restricted to certain conductor models and we do not suppose that the magnetoquasistatic assumption holds. We deduce that the index of the coupled system depends on the chosen gauge. For the coupled electromagnetic device/circuit model using Lorenz gauge we extend the topological index conditions for the modified nodal analysis. The Coulomb gauge always results in an index-2 differential-algebraic equation.

All considered differential-algebraic equations from our application areas have a common structure such as a properly stated leading term, constant projectors onto/along certain subspaces, linear index-2 variables and do not exceed index-2. That is, for all resulting differential-algebraic equations we obtain solvability results, perturbation index results and we can determine consistent initial values. In particular, we show that the perturbation index coincides with the tractability index and does not exceed perturbation index-2.

The first chapter is devoted to the description of the differential-algebraic equations with a properly stated leading occurring in this thesis. The analysis is guided by the tractability index concept up to index-2. We investigate index reduction, solvability and perturbation results. Methods for computing consistent initializations are derived. The following chapter introduces the fundamentals of Maxwell's equations using a potential formulation and discusses boundary and different gauge conditions. We briefly introduce the finite integration technique for spatial discretization. The structural properties of the formulated differential-algebraic equations with incorporated boundary conditions are discussed and we introduce a new class of gauge conditions formulated for the finite integration technique. Index results using different gauge conditions are derived. Chapter 3 is devoted to a detailed network analysis. The modified nodal analysis including memristor models is derived and new topological index criteria and structural properties of the resulting differential-algebraic equations are deduced. In chapter 4 we investigate the coupled system consisting of circuits refined by spatially resolved electromagnetic devices modeled by the modified nodal analysis and Maxwell's equations. We generalize the topological index criteria for the modified nodal analysis for this coupled system and present its structural properties. The final chapter provides proof-of-concept examples to verify the different models including memristors and electromagnetic devices. Appendix A and B subsume the basic aspects of linear algebra and graph theory relevant for this thesis. Appendix C collects auxiliary calculations needed to prove the index results.

2 Differential-Algebraic Equations

Differential/Algebraic Equations are not ODE's.

LINDA PETZOLD, [PET82]

Two major application fields of differential-algebraic equations are the simulation of electric networks and constrained systems. These application areas in engineering can be seen as important impulse to start with a systematic differential-algebraic equation research, since failures in numerical simulations have provoked to analyze these equations.

During the last three decades considerable progress in differential-algebraic equation theory has been made and we refer to [GM86, HLR89, BCP96, ESF98, AP98, HNW02, RR02, KM06, Ria08, LMT13].

Mostly differential-algebraic equations occur because of simplifications of the real problem. In electric networks Kirchhoff current law gives rise to algebraic relationships. If these were modeled either as they are really found or with less idealizations we would obtain an ordinary differential equation or a partial differential equation. In mechanical systems the simple pendulum model has a fixed constraint on the pendulum length whereas any real material will stretch very slightly, see [Gea06].

Differential-algebraic equations are known to be ill-posed in the sense of Hadamard. This ill-posedness is characterized by the differential-algebraic equation index. Briefly speaking, the index can be seen as a measure of the systems' sensitivity to input perturbations, as a measure of the difficulties to be found in the numerical simulation and as the difference to an ordinary differential equation. Depending on the point of view several index definitions exist which mostly generalize the Kronecker index in the linear time-independent case.

In this chapter we introduce the basic notation and tools for the analysis of differential-algebraic equations with a properly stated leading term which occur in this thesis. First, we establish the abstract term of a differential-algebraic equation and point out the main problems we face when dealing with differential-algebraic equations. Second, we briefly introduce some well-known index concepts from literature. Then we familiarize with differential-algebraic equations having a properly stated leading term guided by the tractability index concept. Lastly, we deduce new results in index reduction, local solvability, perturbation results and consistent initialization of differential-algebraic equations.

An implicit ordinary differential equation (ODE - Ordinary Differential Equation) is an equation of the form

$$f\left(\frac{d}{dt}y, y, t\right) = 0, \tag{2.1}$$

where $f \in \mathcal{C}(\mathbb{R}^n \times \mathcal{D} \times \mathcal{I}, \mathbb{R}^n)$ is given and $y \in \mathcal{C}^1(\mathcal{I}, \mathcal{D})$ denotes the unknown function with $\mathcal{D} \subset \mathbb{R}^n$, $\mathcal{I} \subset \mathbb{R}$. In this thesis we restrict ourselves to *initial value problems*, (IVP - Initial-Value Problem) of the form

$$f\left(\frac{d}{dt}y, y, t\right) = 0 \text{ with } y(t_0) = y_0 \in \mathcal{D} \text{ and } t_0 \in \mathcal{I}.$$

Definition 2.1. Let be $(y, t) \in \mathcal{D} \times \mathcal{I}$ with $\mathcal{D} \subset \mathbb{R}^n$ and $\mathcal{I} \subset \mathbb{R}$. We call the implicit ODE (2.1) a *differential-algebraic equation*, (DAE - Differential-Algebraic Equation) if $f \in \mathcal{C}(\mathbb{R}^n \times \mathcal{D} \times \mathcal{I}, \mathbb{R}^n)$, the continuous partial derivatives $\frac{\partial}{\partial y}f(z, y, t)$ and $\frac{\partial}{\partial z}f(z, y, t)$ exist and, in addtion, the partial derivative $\frac{\partial}{\partial z}f(z, y, t)$ is singular with constant rank for all $(z, y, t) \in \mathbb{R}^n \times \mathcal{D} \times \mathcal{I}$.

DAEs have, amongst others, the following two important properties:

- Several components of the solution are determined by constraints. For IVPs these constraints limit the choice of initial values since there is not a solution through every given initial value.

- DAEs with an index higher than two do not only represent integration problems but also differentiation problems. This implies that some parts of the DAE must be differentiable sufficiently often and the differentiations and integrations may be intertwined in a complex manner.

The behavior of DAEs differs from that of explicit ODEs in several aspects. In the following we describe some of the essential differences.

Example 2.2. Regarding the DAE

$$\frac{d}{dt}y_2 = y_2 + y_1 \qquad\qquad y_1 = q(t)$$

with the solution

$$y_1 = q(t)$$

and y_2 being the solution to the explicit ODE $\frac{d}{dt}y_2 = y_2 + q(t)$ for a given input function q. The solution has the properties:

- Only y_2 has to be continuously differentiable with respect to t.

- The initial value for y_1 is fixed by the input function q.

Example 2.3. Regarding the DAE

$$\frac{\mathrm{d}}{\mathrm{dt}}y_1 = y_2 - y_1 \qquad\qquad y_1 = q(t)$$

with the solution

$$y_2 = \frac{\mathrm{d}}{\mathrm{dt}}q(t) + q(t) \qquad\qquad y_1 = q(t)$$

where q is a given input function. The solution has the properties:

- The input function q has to be continuously differentiable with respect to t.

- The initial values are completely fixed by the input function q and $\frac{\mathrm{d}}{\mathrm{dt}}q$.

- To get a solution to y_2 we need to differentiate y_1 with respect to t.

Example 2.4. DAEs are are ill-posed problems. Regarding

$$\frac{\mathrm{d}}{\mathrm{dt}}x_1 = x_2 \quad \text{and} \quad x_1 = \sin(t) + \delta(t)$$

where $\delta(t)$ is a perturbation of the system, the solution is given by

$$x_1 = \sin(t) + \delta(t) \text{ and } x_2 = \cos(t) + \frac{\mathrm{d}}{\mathrm{dt}}\delta(t).$$

A very small perturbation $\delta(t)$, for example $\delta(t) = 10^{-k}\sin(10^{2k}t)$ with $k \gg 1$, can have a serious impact on the solution when compared to the solution $x_2 = \cos(t)$ of the unperturbed problem, where $\delta = 0$, since $\frac{\mathrm{d}}{\mathrm{dt}}\delta(t) = 10^k\cos(10^{2k}t)$.

2.1 Brief Index Survey

> ...and please no war between the different index camps ...
>
> ANDREAS GRIEWANK DURING THE "TWELFTH EUROPEAN WORKSHOP ON AUTOMATIC DIFFERENTIATION WITH EMPHASIS ON APPLICATIONS TO DAES", DEC. 09, 2011, BERLIN.

In this section, we briefly introduce some well-known index concepts from literature namely the Kronecker index, the differential index, the perturbation index and the strangeness index. In the majority of cases DAEs arising in industrial applications are nonlinear. The Kronecker index is only defined for linear DAEs with constant coefficients. Other index definitions are mostly generalizations of the Kronecker index for the time-varying and nonlinear case. The different index definitions depend on various perspectives but all concepts exist in their own right and each has its own pros and cons.

Kronecker Index

The first introduced index concept was the Kronecker index [GP83, GM86]. The concept is only defined for linear DAEs (2.1) with constant coefficients given by

$$A \frac{d}{dt} y + By = q \qquad (2.2)$$

with $A, B \in \mathbb{R}^{n \times n}$, $q \in \mathcal{C}(\mathcal{I}, \mathbb{R}^n)$ and $y \in \mathcal{C}^1(\mathcal{I}, \mathcal{D})$, where A is singular. For this type of DAEs the Kronecker index provides a closed solution formula. The Kronecker index is closely related to regular matrix pencils.

Definition 2.5 ([Gan71]). Let be $A, B \in \mathbb{R}^{n \times n}$. The ordered *matrix pair* $\{A, B\}$ and the *matrix pencil* $\lambda A + B$ respectively are called *nonsingular* if there is a constant $\lambda \in \mathbb{R}$ so that $\det(\lambda A + B) \not\equiv 0$. Otherwise they are called *singular*.

Both the ordered matrix pair $\{A, B\}$ and the linear DAE (2.2) are said to be regular if the accompanying matrix pencil is nonsingular. In fact, the regular matrix pencil is essential for the unique solvability of the DAE (2.2), see [LMT13].

Lemma 2.6. If the matrix pair $\{A, B\}$ is nonsingular, then $\frac{1}{h}A + B$ is nonsingular for sufficiently small $h > 0$.

Proof. We regard the polynomial $\det(\lambda A + B)$ in λ. If $\det(\lambda A + B) \not\equiv 0$ then there is only a finite number of roots of the polynomial. Let λ_0 be the root with the largest absolute value. Then $\frac{1}{h}A + B$ is nonsingular for all $0 < h < \frac{1}{|\lambda_0|}$. $\qquad \square$

Theorem 2.7. For any regular matrix pair $\{A, B\}$ there are nonsingular matrices $L, K \in \mathbb{R}^{n \times n}$ and an integer $0 \leqslant l \leqslant n$ such that

$$LAK = \begin{bmatrix} I & 0 \\ 0 & N \end{bmatrix} \text{ and } LBK = \begin{bmatrix} W & 0 \\ 0 & I \end{bmatrix} \qquad (2.3)$$

with $N \in \mathbb{R}^{l \times l}$ and $W \in \mathbb{R}^{(n-l) \times (n-l)}$. Here, N is absent if $l = 0$. Otherwise there is $0 \leqslant k \leqslant l$ such that N is nilpotent of order k, that is, $N^k = 0$ and $N^{k-1} \neq 0$. The integers l and k as well as the eigenstructure of the blocks N and W are uniquely determined by the matrix pair $\{A, B\}$.

Proof. See Proposition 1.3 in [LMT13]. $\qquad \square$

The matrix N in Theorem 2.7 has only the eigenvalue zero and can be transformed into its Jordan normal form by means of a real valued similarity transformation. Therefore, the transformation matrices L and K can be chosen such that N is in Jordan form. The pair given by (2.3) is called Weierstraß-Kronecker form of the regular matrix pair $\{A, B\}$, see [Gan71].

Definition 2.8 (*Kronecker index*). The Kronecker index of a matrix regular pair $\{A, B\}$ and the Kronecker index of a regular DAE (2.2) are defined to be the nilpotency order k in the Weierstraß-Kronecker form (2.3).

The DAE (2.2) in Weierstraß-Kronecker form is completely decoupled and provides a broad insight into the structure of the DAE. Every regular DAE (2.2) with Kronecker index-k can be transformed into the Weierstraß-Kronecker form given by

$$\frac{d}{dt}u + Wu = p \tag{2.4}$$

$$N\frac{d}{dt}v + v = r \tag{2.5}$$

with $y = K \begin{pmatrix} u \\ v \end{pmatrix}$ and $Lq = \begin{pmatrix} p \\ r \end{pmatrix}$. We obtain the explicit ODE (2.4) and for $l > 0$ we deduce from (2.5) the unique solution

$$v = \sum_{i=0}^{k-1} (-1)^i N^i r^{(i)}, \tag{2.6}$$

provided $r \in \mathcal{C}^{k-1}(\mathcal{I}, \mathbb{R})$ by recursive use of (2.5), see [LMT13]. Thus (2.6) shows the dependence of the solution y on derivatives of the right-hand side or the perturbation term q.

The higher the index the more differentiations are needed. From the numerical point of view it is very important to know the index of a DAE (2.2) as well as details on the structure responsible for differentiations. The regularity of the matrix pair $\{A, B\}$ guarantees the unique solvability for linear constant DAE (2.2) if we assume smooth input functions q. If A and B are time-dependent this unfortunately holds no longer true. There are examples, where the matrix pair is regular for all $t \in \mathcal{I}$ and the DAE has infinitely many solutions. It may also happen that the matrix pair is singular and a unique solution exists, see [BCP96].

If $r \in \mathcal{C}^k(\mathcal{I}, \mathbb{R})$, the differentiation of (2.6) yields an ODE for v. That idea is picked up by the differentiation index that figures out how many differentiations are necessary to transform the DAE (2.2) into an ODE. On the other hand the perturbation index directly measures the impact of perturbations on the solution.

Differentiation Index

The best known index is probably the differentiation index [Cam87, BCP96]. It is more or less the number of differentiations needed to transform a DAE into an ODE. The differentiation index received much attention and it is widely used. But it assumes high smoothness of the DAE which often does not hold for applications.

Definition 2.9 (*differentiation index*). The DAE (2.1) has differentiation index-k if $f \in \mathcal{C}^k(\mathbb{R}^n \times \mathcal{D} \times \mathcal{I}, \mathbb{R}^n)$ and k is the minimal number of analytical differentiations with respect to t needed to determine an ODE for $\frac{d}{dt}y$ as a continuous function in y and t by algebraic manipulations only.

A major drawback in application of the differentiation index is that the calculated differentiation index-k is just an upper bound for the exact differentiation index of the system and the exact index can be lower than k. The calculated differentiation index depends strongly on a successful rearranging of the system's unknowns. However, the differentiation index is not clearly defined for DAEs as

$$y \frac{\mathrm{d}}{\mathrm{d}t} y = 0.$$

Here $y \equiv 0$ is a solution, but then we can not determine an ODE for y. Otherwise for $y \equiv 1$ we obtain $\frac{\mathrm{d}}{\mathrm{d}t}y = 0$ and hence differentiation index-1.

Perturbation Index

The perturbation index [HLR89, HNW02] interprets the index as a measure of sensitivity of the solution with respect to perturbations of the given problem and the right hand side. The perturbation may arise from rounding errors and numerical approximations. From a numerical point of view the perturbation index is the most important one. A major drawback is that the perturbation index does not give us a prescription way how to determine it and requires knowledge about the exact solution.

Definition 2.10 (*perturbation index*). The DAE (2.1) has perturbation index-k along a solution $y_* \in \mathcal{C}^1 (\mathcal{I}, \mathcal{D})$ on a compact interval $\mathcal{I} = [t_0, T]$, if k is the smallest number so that for all functions $\bar{y} \in \mathcal{C}^1 (\mathcal{I}, \mathcal{D})$ with

$$f \left(\frac{\mathrm{d}}{\mathrm{d}t} \bar{y}, \bar{y}, t \right) = q(t)$$

for $q \in \mathcal{C}^{k-1} (\mathcal{I}, \mathbb{R}^n)$ and all $t \in \mathcal{I}$, the inequality

$$\|\bar{y} - y_*\|_\infty \leqslant c \left(\|\bar{y}(t_0) - y_*(t_0)\|_\infty + \sum_{j=0}^{k-1} \|q^{(j)}\|_\infty \right)$$

holds true for some $c > 0$ as long as $\|q^{(j)}\|_\infty$, $j < k$, and $\|\bar{y}(t_0) - y_*(t_0)\|_\infty$ are sufficiently small.

Strangeness Index

The strangeness index [KM94, KM06] is an algorithmic approach, relies on a transformation to a canonical form and is closely related to the differentiation index, but extended to over- and under-determined systems.

We will briefly review some of the key properties related to the strangeness index, see [Voi06]. From the Kronecker index point of view two matrix pairs $\{A_1, B_1\}$ and $\{A_2, B_2\}$ are considered to be equivalent if there exist nonsingular matrices U and V, so that

$$A_2 = UA_1V \text{ and } B_2 = UB_1V.$$

For $A_1 = A$ and $B_1 = B$ the DAE (2.2) is rewritten in terms of the transformed unknown $x = V^{-1}y$. If V depends on the time, then $y = Vx$ needs to be differentiated to obtain the transformed DAE. Thus, $\frac{d}{dt}y = \frac{d}{dt}Vx + V\frac{d}{dt}x$ holds true and the additional term $\frac{d}{dt}Vx$ has to be taken into account.

The strangeness index concept considers two time-dependent matrix pairs $\{A_1, B_1\}$ and $\{A_2, B_2\}$, with $A_i, B_i \in \mathcal{C}(\mathbb{R}, \mathbb{R}^{n\times m})$ and $i = 1, 2$, equivalent, if there exist point-wise nonsingular matrix functions $U \in \mathcal{C}(\mathbb{R}, \mathbb{R}^{n\times n})$ and $V \in \mathcal{C}(\mathbb{R}, \mathbb{R}^{m\times m})$, so that

$$A_2 = UA_1V \text{ and } B_2 = UB_1V + UA_1\frac{d}{dt}V.$$

Under certain constant rank assumptions it is possible to derive the normal form

$$\{A_1, B_1\} \sim \{A_2, B_2\} = \left\{ \begin{bmatrix} I_s & 0 & 0 & 0 \\ 0 & I_d & 0 & 0 \\ 0 & 0 & 0 & 0 \\ 0 & 0 & 0 & 0 \\ 0 & 0 & 0 & 0 \end{bmatrix}, \begin{bmatrix} 0 & A_{12} & 0 & A_{14} \\ 0 & 0 & 0 & A_{24} \\ 0 & 0 & I_a & 0 \\ I_s & 0 & 0 & 0 \\ 0 & 0 & 0 & 0 \end{bmatrix} \right\} \begin{matrix} s_1 \\ d_1 \\ a_1 \\ s_1 \\ m_1 - d_1 - a_1 \end{matrix}$$

where the blocks

$$A_{12} \in \mathcal{C}\left(\mathbb{R}, \mathbb{R}^{s_1 \times d_1}\right), A_{14} \in \mathcal{C}\left(\mathbb{R}, \mathbb{R}^{s_1 \times (m-m_1-d_1-a_1)}\right) \text{ and } A_{24} \in \mathcal{C}\left(\mathbb{R}, \mathbb{R}^{d_1 \times (m-m_1-d_1-a_1)}\right)$$

are again matrix functions. The numbers s_1, d_1 and a_1 are invariants of the equivalence relations, see [KM06]. The corresponding linear DAE (2.2) is found to be equivalent to the DAE:

$$\frac{d}{dt}y_1 + A_{12}y_2 + A_{14}y_4 = q_1 \tag{2.7a}$$

$$\frac{d}{dt}y_2 + A_{24}y_4 = q_2 \quad \text{(dynamic part)}$$

$$y_3 = q_3 \quad \text{(algebraic part)}$$

$$y_1 = q_4 \tag{2.7b}$$

$$0 = q_5 \quad \text{(consistency condition)}$$

The "strangeness" is derived from the coupling between (2.7a) and (2.7b). Differentiating (2.7b) and inserting into (2.7a) leads to an algebraic equation and we get the DAE:

$$A_{12}y_2 + A_{14}y_4 = q_1 - \frac{d}{dt}q_4$$

$$\frac{d}{dt}y_2 + A_{24}y_4 = q_2$$

$$y_3 = q_3$$

$$y_1 = q_4$$

$$0 = q_5$$

The resulting modified matrix pair is again denoted by $\{A_2, B_2\}$. The procedure to obtain a normal from and the elimination of the "strangeness" can be repeated to obtain a sequence of characteristic values (s_i, d_i, a_i) for the matrix pairs $\{A_i, B_i\}$.

A matrix pair $\{A_i, B_i\}$ is called strangeness-free if $s_i = 0$. The *strangeness index* is i, if $i \in \mathbb{N}$ is the smallest number so that the matrix pair $\{A_i, B_i\}$ is strangeness-free, see [KM06].

The strangeness index is a powerful tool for the analysis of DAEs, including over- and under-determined systems. The resulting normal forms provide much inside into the structure of a given DAE. But even for simple DAEs it may be difficult to calculate the normal forms, in particular for nonlinear problems.

Other Index Concepts

> Thanks to Caren we have all the numerical problems.
>
> ANDREAS STEINBRECHER DURING THE "TWELFTH EUROPEAN WORKSHOP ON AUTOMATIC DIFFERENTIATION WITH EMPHASIS ON APPLICATIONS TO DAEs" AS RESPONSE TO CAREN TISCHENDORF'S EXAMPLE POINTING OUT A GAP IN THE STRUCTURAL INDEX CONCEPT, DEC. 9, 2011, BERLIN.

In addition to the index concepts already mentioned a geometrical theory to study DAEs as ODEs on manifolds is provided by the geometrical index [RR90, RR02]. A combinatorial index concept is the structural index [BMR00, Pry01]. In [Jan12a] a new index concept is introduced combining the ideas of the tractability index and the strangeness index. For more index discussions we refer to [Voi06, Meh12].

Expect the structural index all index concepts mentioned are generalizations of the Kronecker index in case of linear DAEs (2.2) with constant coefficients.

2.2 Analysis of Differential-Algebraic Equations

> Actually, **DAEs ARE ODEs** but those which cannot be solved with respect to x'.
>
> EBERHARD GRIEPENTROG, MICHAEL HANKE AND ROSWITHA MÄRZ, [GHM92]

Apart from the index concepts discussed so far there is the tractability index concept. The tractability index is a projector-based algorithmic decoupling concept working in terms of the original unknowns and it is straightforward to determine the tractability index at least in theory. The concept behind the tractability index is a stepwise projection of the solution onto certain invariant subspaces leading to a precise solution description, see [LMT13]. It focuses on the linearization of a DAE and requires only weak smoothness conditions. The decoupling procedure provides a detailed insight into the structure of a given DAE, see [GM86, Ria08, LMT13]. In addition for some classes of DAEs we can connect the tractability index with the perturbation index, which is of importance for our purposes.

Quasilinear DAEs (2.1) can be written as

$$A\left(y, t\right) \frac{d}{dt} y + b\left(y, t\right) = 0,$$

where $A \in \mathcal{C}\left(\mathcal{D} \times \mathcal{I}, \mathbb{R}^{n \times n}\right)$ and $b \in \mathcal{C}\left(\mathcal{D} \times \mathcal{I}, \mathbb{R}^n\right)$. Formally, it must be assumed that the solution $y \in \mathcal{C}^1\left(\mathcal{I}, \mathcal{D}\right)$ is more smooth than actually required since $A\left(y, t\right)$ is singular and hence only the solution components in the cokernel of $A\left(y, t\right)$ have to be continuously differentiable. To avoid the unnecessary smoothness we focus on DAEs (2.1) of the more special form

$$A\left(y, t\right) \frac{d}{dt} d\left(y, t\right) + b\left(y, t\right) = 0 \tag{2.8}$$

with

- $A \in \mathcal{C}\left(\mathcal{D} \times \mathcal{I}, \mathbb{R}^{n \times m}\right)$, $d \in \mathcal{C}^1\left(\mathcal{I}, \mathbb{R}^m\right)$, $b \in \mathcal{C}\left(\mathcal{D} \times \mathcal{I}, \mathbb{R}^n\right)$,

- a properly stated leading term, see Definition 2.11,

and

- the continuous partial derivatives $\frac{\partial}{\partial y} A\left(y, t\right)$, $\frac{\partial}{\partial y} d\left(y, t\right)$, $\frac{\partial}{\partial t} d\left(y, t\right)$ and $\frac{\partial}{\partial y} b\left(y, t\right)$ exist.

We denote $D\left(y, t\right) = \frac{\partial}{\partial y} d\left(y, t\right)$. The *leading term* $d\left(y, t\right)$ figures out precisely which derivatives are actually involved and we need the continuous differentiability of combinations of the solution components only in the cokernel of $A\left(y, t\right)$.

Definition 2.11 ([Mär01]). The DAE (2.8) has a *properly stated leading term* if a projector $R \in \mathcal{C}^1\left(\mathcal{I}, \mathbb{R}^{m \times m}\right)$ exists with

$$\ker A\left(y, t\right) = \ker R\left(t\right) \text{ and } \operatorname{im} D\left(y, t\right) = \operatorname{im} R\left(t\right)$$

for all $\left(y, t\right) \in \mathcal{D} \times \mathcal{I}$.

Hence $\ker A\left(y, t\right)$ and $\operatorname{im} D\left(y, t\right)$ do not depend on $y \in \mathcal{D}$ for a DAE with a properly stated leading term and the subspaces have constant dimensions. Furthermore they are well matched together without any overlap or gap. Once again the leading term shows precisely all involved derivatives. Due to the historical development most DAEs are formulated without a properly stated leading term.

Lemma 2.12. For the DAE (2.8)

$$\ker A\,(y,t) \oplus \operatorname{im} D\,(y,t) = \mathbb{R}^m \tag{2.9}$$

holds true, which is equivalent to

$$\operatorname{im} A\,(y,t) = \operatorname{im} A\,(y,t)\,D\,(y,t) \text{ and } \ker D\,(y,t) = \ker A\,(y,t)\,D\,(y,t)\,.$$

In addition the identities $A\,(y,t)\,R\,(t) = A\,(y,t)$ and $R\,(t)\,D\,(y,t) = D\,(y,t)$ hold true for all $(y,t) \in \mathcal{D} \times \mathcal{I}$.

Proof. See Lemma A.1.3 in [LMT13] and Lemma A.9. $\qquad\qquad\square$

Definition 2.13. A function $y \in \mathcal{C}\,(\mathcal{I}, \mathbb{R}^n)$ is said to be a *solution* to the DAE (2.8) if $y \in \mathcal{C}_d^1\,(\mathcal{I}, \mathcal{D})$ with the *canonical solution set*

$$\mathcal{C}_d^1\,(\mathcal{I}, \mathcal{D}) = \left\{ y \in \mathcal{C}\,(\mathcal{I}, \mathcal{D}) \mid d\,(y\,(\cdot), \cdot) \in \mathcal{C}^1\,(\mathcal{I}, \mathbb{R}^m) \right\}$$

and the DAE (2.8) is fulfilled pointwisely.

The solution set of the DAE (2.8) is nonlinear if $d\,(y,t)$ is nonlinear with respect to y. Fortunately it is straightforward to transform the DAE (2.8) into a DAE of the form

$$\overline{A}\,(\overline{y},t)\,\frac{d}{dt}\left[\overline{D}\,(t)\,\overline{y}\right] + \overline{b}\,(\overline{y},t) = 0 \tag{2.10}$$

with a properly stated leading term as shown in the following. The transformation to a DAE (2.10) makes useful implication such as a solvability and perturbation result available for the DAE (2.8) for a certain DAE class as we will see in the next section.

Definition 2.14. A function $y \in \mathcal{C}\,(\mathcal{I}, \mathbb{R}^n)$ is said to be a solution to the DAE (2.10) if $\overline{y} \in \mathcal{C}_{\overline{D}}^1\,(\mathcal{I}, \mathcal{D})$ with the canonical *linear solution space*

$$\mathcal{C}_{\overline{D}}^1\,(\mathcal{I}, \mathcal{D}) = \left\{ \overline{y} \in \mathcal{C}\,(\mathcal{I}, \mathcal{D}) \mid \overline{D}\,(\cdot)\,\overline{y}\,(\cdot) \in \mathcal{C}^1\,(\mathcal{I}, \mathbb{R}^m) \right\}$$

and the DAE (2.10) is fulfilled pointwisely.

This allows linearization of the DAE which based on linear function spaces, see [Mär01]. In fact, $\mathcal{C}_{\overline{D}}^1\,(\mathcal{I}, \mathcal{D})$ is a vector space over $\mathbb{R}$ using point-wise addition and scalar multiplication. Together with the norm

$$\|\overline{y}\|_{\mathcal{C}_{\overline{D}}^1} = \|\overline{y}\|_\infty + \left\| \frac{d}{dt}\left[\overline{D}\,(t)\,\overline{y}\right] \right\|_\infty$$

we obtain the Banach space $\left(\mathcal{C}_{\overline{D}}^1, \|\cdot\|_{\mathcal{C}_{\overline{D}}^1} \right)$, see Theorem 9 in [GM86].

Definition 2.15. The *natural extension* of a DAE (2.8) is given by

$$\overline{A}\,(\overline{y},t)\,\frac{d}{dt}\left[\overline{D}\,(t)\,\overline{y}\right] + \overline{b}\,(\overline{y},t) = 0 \tag{2.11}$$

with

$$\overline{y} = \begin{bmatrix} y \\ z \end{bmatrix},\ \overline{A}\,(\overline{y},t) = \begin{bmatrix} A\,(y,t) \\ 0 \end{bmatrix},\ \overline{D}\,(t) = \begin{bmatrix} 0 & R\,(t) \end{bmatrix}\ \text{and}\ \overline{b}\,(\overline{y},t) = \begin{bmatrix} b\,(y,t) \\ z - d\,(y,t) \end{bmatrix}.$$

The original DAE (2.8) is called the underlying DAE to the natural extension.

The natural extension and the underlying DAE are closely related as shown in the next theorem.

Theorem 2.16 ([Mär01]). The natural extension (2.11)

(i) is a DAE of the form (2.10)

(ii) and the underlying DAE are equivalent by the relation $z = d\,(y,t)$, $t \in \mathcal{I}$.

Proof. (i) Clearly $\ker A\,(y,t) = \ker \overline{A}\,(\overline{y},t)$ and $\operatorname{im} D\,(y,t) = \operatorname{im} \overline{D}\,(t)$ due to the properly stated leading term of the underlying DAE (2.8). Hence we can choose $\overline{R}\,(t) = R\,(t)$ and the natural extension (2.11) has a properly stated leading term as well.
(ii) If $y_* \in \mathcal{C}_d^1\,(\mathcal{I},\mathcal{D})$ is a solution to the DAE (2.8) then $\overline{y}_* \in \mathcal{C}_{\overline{D}}^1\,(\mathcal{I},\mathcal{D} \times \mathbb{R}^n)$ with $z = d\,(y,t)$, $t \in \mathcal{I}$, is a solution to the natural extension (2.11). If $\overline{y}_* \in \mathcal{C}_{\overline{D}}^1\,(\mathcal{I},\mathcal{D} \times \mathbb{R}^n)$ is a solution to the natural extension (2.11), then $d\,(y,t) = R\,(t)\,z \in \mathcal{C}^1\,(\mathcal{I},\mathbb{R}^m)$ holds true and hence $y_* \in \mathcal{C}_d^1\,(\mathcal{I},\mathcal{D})$ is a solution to the underlying DAE (2.8). □

Obviously if $y \in \mathcal{C}\,(\mathcal{I},\mathbb{R}^n)$ is a solution to the DAE (2.8) then $y\,(t) \in \mathcal{M}_0\,(t)$ for all $t \in \mathcal{I}$ must hold true with

$$\mathcal{M}_0\,(t) = \{y\,(t) \in \mathcal{D} \mid b\,(y,t) \in \operatorname{im} A\,(y,t)\} \subset \mathbb{R}^n$$

is the so-called *obvious constraint set*, see [Mär03]. The flow of the DAE (2.8) is restricted to $\mathcal{M}_0\,(t)$ and there is no solution through every given initial value. Thus, for the numerical integration of the DAE (2.8) we need to start the integration using a suitable intial value.

Definition 2.17. A value $y_0 \in \mathcal{M}_0\,(t_0)$ is said to be a *consistent initial value* of the DAE (2.8) if there is a solution passing through $(y_0,t_0) \in \mathcal{D} \times \mathcal{I}$.

Definition 2.18. A triple $(z_0,y_0,t_0) \in \mathbb{R}^m \times \mathcal{M}_0\,(t_0) \times \mathcal{I}$ is said to be an *operating point* of the DAE (2.8) if

$$A\,(y_0,t_0)\,z_0 + b\,(y_0,t_0) = 0$$

is fulfilled.

The term operating point comes originally from circuit simulation, which is an important application class for this thesis. In some cases, depending on the integration method, we are also interested in a starting value of the derivatives appearing in the DAE (2.8). The following lemma and definition will characterize the values of the derivatives properly.

Lemma 2.19. For every $y_0 \in \mathcal{M}_0(t_0)$, $t_0 \in \mathcal{I}$, of the DAE (2.8) there is a unique $z_0 \in \mathbb{R}^m$ such that

$$A(y_0, t_0) z_0 + b(y_0, t_0) = 0 \text{ and } z_0 = R(t_0) z_0 \qquad (2.12)$$

are fulfilled.

Proof. Let $y_0 \in \mathcal{M}_0(t_0)$, $t_0 \in \mathcal{I}$ and $z_0^1, z_0^2 \in \mathbb{R}^m$ be fulfilling (2.12). Then

$$A(y_0, t_0) \left(z_0^1 - z_0^2 \right) = 0 \text{ and } z_0^1 - z_0^2 \in \ker R(t_0).$$

Furthermore $z_0^1 - z_0^2 = R(t_0) (z_0^1 - z_0^2)$ and $z_0^1 - z_0^2 \in \operatorname{im} R(t_0)$ are valid. Hence we have $z_0^1 - z_0^2 = 0$. $\qquad \square$

Definition 2.20. A triple $(z_0, y_0, t_0) \in \operatorname{im} R(t_0) \times \mathcal{M}_0(t_0) \times \mathcal{I}$ is said to be a *consistent initialization* of the DAE (2.8) if $y_0 \in \mathcal{M}_0(t_0)$ is a consistent value and the triple is an operating point.

For our investigations we choose the tractability index concept. Once again the tractability index is a projector-based algorithmic decoupling concept and the concept behind it is a stepwise projection of the solution onto certain invariant subspaces, see [LMT13]. In this thesis we focus on DAEs (2.8) of tractability index-1 and index-2 which occur in our applications. Next we define the needed matrices and subspaces for this index concept.

Definition 2.21 (*Matrix Chain* and Subspaces). Given the DAE (2.8) we define:

$$G_0(y, t) = A(y, t) D(y, t)$$

$$B_0(z, y, t) = \frac{\partial}{\partial y} \left[A(y, t) z + b(y, t) \right]$$

$$P_0(y, t) = I - Q_0(y, t), \ Q_0(y, t) \text{ projector onto } \ker G_0(y, t)$$

$$\mathcal{N}_0(y, t) = \ker G_0(y, t)$$

$$\mathcal{S}_0(z, y, t) = \{v \in \mathbb{R}^n | B_0(z, y, t) v \in \operatorname{im} G_0(y, t)\}$$

$$G_1(z, y, t) = G_0(y, t) + B_0(z, y, t) Q_0(y, t)$$

$$P_1(z, y, t) = I - Q_1(z, y, t), \ Q_1(z, y, t) \text{ projector onto } \ker G_1(z, y, t)$$

$$\mathcal{N}_1(z, y, t) = \ker G_1(z, y, t)$$

$$\mathcal{S}_1(z, y, t) = \{v \in \mathbb{R}^n | B_0(z, y, t) P_0(y, t) v \in \operatorname{im} G_1(y, t)\}$$

$$G_2(z, y, t) = G_1(z, y, t) + B_0(z, y, t) P_0(y, t) Q_1(z, y, t)$$

We choose the projector $Q_1(z, y, t)$ such that $\mathcal{N}_0(y, t) \subset \ker Q_1(z, y, t)$.

Remark 2.22. The choice of the projector $Q_1(z, y, t)$ so that $\mathcal{N}_0(y, t) \subset \ker Q_1(z, y, t)$ is always possible. The matrix chain is said to be admissible up to two, see [LMT13].

For computational aspects of the matrix chain as well as for the properly stated leading term we refer to [LMT13] taking Remark 2.31 into account.

Definition 2.23 (*tractability index*). The DAE (2.8) has (tractability)

- index-0 if and only if the *index-0 set* $\mathcal{N}_0(y, t)$ satisfies

$$\mathcal{N}_0(y, t) = \{0\}$$

- index-1 if and only if the DAE (2.8) does not have index-0 and the *index-1 set* $\mathcal{N}_0(y, t) \cap \mathcal{S}_0(z, y, t)$ satisfies

$$\mathcal{N}_0(y, t) \cap \mathcal{S}_0(z, y, t) = \{0\}$$

- index-2 if and only if the DAE (2.8) has neither index-0 nor index-1, the index-1 set satisfies

$$\dim(\mathcal{N}_0(y, t) \cap \mathcal{S}_0(z, y, t)) = \text{const.}$$

and the *index-2 set* $\mathcal{N}_1(y, t) \cap \mathcal{S}_1(z, y, t)$ satisfies

$$(\mathcal{N}_1 \cap \mathcal{S}_1)(z, y, t) = \{0\}$$

for all $(z, y, t) \in \mathbb{R}^m \times \mathcal{D} \times \mathcal{I}$.

Remark 2.24 ([Mär02]). For the matrix chain of the DAE (2.8) holds:

(i) The projectors $Q_0(y, t)$ and $Q_1(z, y, t)$ are not uniquely determined.

(ii) If $\mathcal{N}_0(y, t) \subset \ker Q_1(z, y, t)$, then $Q_1(z, y, t) Q_0(y, t) = 0$.

(iii) The index is independent of the choice of the projectors $Q_0(y, t)$ and $Q_1(z, y, t)$ as long as (ii) is valid.

Furthermore the index of the DAE (2.8) is invariant under transformation and scaling.

It is worth to formulate a DAE with a properly stated leading term because for a large class of index-1 and index-2 DAEs it has been shown that backward differentiation formulas (BDF - Backward Differentiation Formulas) and Runge-Kutta methods are stability preserving, see [HMT03a, HMT03b]. Such a DAE formulation is called *numerically qualified*. An appropriate formulation of the problem ensures a correct behavior of the numerical solution. The numerical methods keep their stability properties and unexpected step size restrictions can be avoided.

Example 2.25. Consider the linear index-1 DAE

$$\begin{bmatrix} \lambda t & \lambda - 1 \\ 0 & 0 \end{bmatrix} \frac{d}{dt} \begin{pmatrix} u \\ v \end{pmatrix} + \begin{bmatrix} 0 & 0 \\ \lambda t - 1 & \lambda - 1 \end{bmatrix} \begin{pmatrix} u \\ v \end{pmatrix} = 0 \tag{2.13}$$

reported [But03] with $\lambda \neq 1$. From the DAE (2.13) we obtain the ODE $\frac{d}{dt}u = \lambda u$. Using the implicit Euler to solve the DAE (2.13) we obtain

$$u_{n+1} = (1 + h\lambda)\, u_n$$

which is in fact the explicit Euler applied to the ODE. This will have several consequences such as step size restrictions due to stability requirements. Formulating the DAE (2.13) with a properly stated leading term may lead to

$$\begin{bmatrix} 1 \\ 0 \end{bmatrix} \frac{d}{dt} \left(\begin{bmatrix} \lambda t & \lambda - 1 \end{bmatrix} \begin{pmatrix} u \\ v \end{pmatrix} \right) + \begin{bmatrix} -\lambda & 0 \\ \lambda t - 1 & \lambda - 1 \end{bmatrix} \begin{pmatrix} u \\ v \end{pmatrix} = 0. \tag{2.14}$$

Using the implicit Euler to solve the DAE (2.14) with a properly stated leading term we obtain

$$u_{n+1} = \frac{1}{1 - h\lambda} u_n$$

and induce the implicit Euler for the ODE, too.

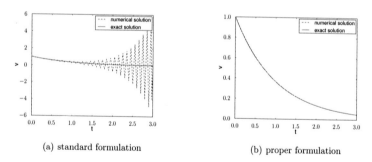

(a) standard formulation   (b) proper formulation

Figure 2.1: BDF-2 solution to the η-DAE with step size $h = 0.01$ and $\eta = -0.275$.

Example 2.26. The well-known linear index-2 DAE, so-called the η-DAE, described in [GP83], further investigated in [HLR89] and successfully tackled by the properly stated leading term in [HMT03b], is given by

$$\begin{bmatrix} 0 & 0 \\ 1 & \eta t \end{bmatrix} \frac{d}{dt} \begin{pmatrix} u \\ v \end{pmatrix} + \begin{bmatrix} 1 & \eta t \\ 0 & 1 + \eta \end{bmatrix} \begin{pmatrix} u \\ v \end{pmatrix} = \begin{pmatrix} e^{-t} \\ 0 \end{pmatrix}$$

with the exact solution $u(t) = (1 - \eta t) e^{-t}$ and $v(t) = e^{-t}$ such that $(u_0, v_0) = (1, 1)$ is a consistent initial value at $t_0 = 0$. Using the original formulation the implicit Euler fails completely for $\eta = -1$ and is exponentially unstable for $\eta < -\frac{1}{2}$, see [GP84]. The simple reformulation

$$\begin{bmatrix} 0 \\ 1 \end{bmatrix} \frac{d}{dt} \left(\begin{bmatrix} 1 & \eta t \end{bmatrix} u \right) + \begin{bmatrix} 1 & \eta t \\ 0 & 1 \end{bmatrix} \begin{pmatrix} u \\ v \end{pmatrix} = \begin{pmatrix} e^{-t} \\ 0 \end{pmatrix}$$

with a properly stated leading term leads to a correct implicit Euler solution, see [HMT03b]. These statements are confirmed by the numerical results given in Figure 2.1.

Remark 2.27. Let $W_0(y, t)$ be a projector along $\operatorname{im} G_0(y, t)$. Then

$$\mathcal{S}_0(z, y, t) = \ker W_0(y, t) B_0(z, y, t)$$

holds true for all $(z, y, t) \in \mathbb{R}^m \times \mathcal{D} \times \mathcal{I}$.

Remark 2.28. Let $W_1(z, y, t)$ be a projector along $\operatorname{im} G_1(z, y, t)$. Then

$$\mathcal{S}_1(z, y, t) = \ker W_1(z, y, t) B_0(z, y, t) P_0(y, t)$$

holds true for all $(z, y, t) \in \mathbb{R}^m \times \mathcal{D} \times \mathcal{I}$.

Now we come back to the relation between the natural extension and their underlying DAE. The index of the natural extension (2.11) is given by the underlying DAE and vice versa.

Theorem 2.29. The natural extension (2.11) and the underlying DAE have both index-1 or index-2.

Proof. See Theorem 3.4 in [Mär01]. $\qquad\qquad\qquad\qquad\qquad\qquad\qquad\square$

At a later stage we will utilize an equivalent characterization for the tractability index. We use this equivalence to prove that the index reduction introduced in the next section really reduce the index.

Lemma 2.30. The DAE (2.8) has (tractability)

- index-0 if and only if $G_0(y, t)$ is nonsingular

- index-1 if and only if the DAE does not have index-0 and $G_1(z, y, t)$ is nonsingular

- index-2 if and only if the DAE has neither index-0 nor index-1 and $G_2(z, y, t)$ is nonsingular

for all $(z, y, t) \in \mathbb{R}^m \times \mathcal{D} \times \mathcal{I}$ with constant rank.

Proof. See [GM86] and [Est00]. $\qquad\qquad\qquad\qquad\qquad\qquad\qquad\qquad\qquad\square$

Remark 2.31. Notice that we define $G_2(z, y, t)$ different to [LMT13]. However, since

$$\overline{G_2}(z, y, t) = G_2(z, y, t)(I - P_1(z, y, t) E(z, y, t) Q_1(z, y, t))$$

holds true with $(I - P_1(z, y, t) E(z, y, t) Q_1(z, y, t))$ nonsingular, see Lemma A.5, it is sufficient to check whether $G_2(z, y, t)$ is nonsingular or not in the index-2 case. For $\overline{G_2}(z, y, t)$ we have $E(z, y, t) = D(y, t)^- \frac{d}{dt}(D(y, t) P_0(z, y, t) P_1(y, t) D(y, t)^-) D(y, t)$.

The next lemma is motivated by [Sch03]. We need the lemma to prove Theorem 2.59. In fact the next lemma provides a decoupling into certain solution components as shown later.

Lemma 2.32. Let the index-2 DAE (2.8) be given. Then

$$G_2(z, y, t)^{-1} G_0(y, t) = P_1(z, y, t) P_0(y, t)$$

and

$$G_2(z, y, t)^{-1} B_0(z, y, t) = G_2(z, y, t)^{-1} B_0(z, y, t) P_0(y, t) P_1(z, y, t)$$
$$+ Q_1(z, y, t) + Q_0(y, t)$$

holds true for all $(z, y, t) \in \mathbb{R}^m \times \mathcal{D} \times \mathcal{I}$.

Proof. The first identity is true since

$$\begin{aligned}
G_0(y, t) &= (G_0(y, t) + B_0(z, y, t) Q_0(y, t)) P_0(y, t) \\
&= G_1(z, y, t) P_0(y, t) \\
&= (G_1(z, y, t) + B_0(z, y, t) P_0(y, t) Q_1(z, y, t)) P_1(z, y, t) P_0(y, t) \\
&= G_2(z, y, t) P_1(z, y, t) P_0(y, t)
\end{aligned}$$

and the second one due to

$$\begin{aligned}
B_0(z, y, t) &= B_0(z, y, t) P_0(y, t) P_1(z, y, t) + B_0(z, y, t) P_0(y, t) Q_1(z, y, t) \\
&\quad + B_0(z, y, t) Q_0(y, t) \\
&= B_0(z, y, t) P_0(y, t) P_1(z, y, t) + G_2(z, y, t) Q_1(z, y, t) + G_2(z, y, t) Q_0(y, t)
\end{aligned}$$

is valid. $\qquad\square$

2.2.1 Index Reduction, Solvability and Perturbation Results

Our goal is to describe all constraint sets for index-2 DAEs with a properly stated leading term having a special structure and to derive a solvability and perturbation result. For this, a suitable tool is the *index reduction*. The index reduction may be applied to a DAE to lower the index down from an initially higher index. A well known approach is the differentiation of the DAE or of parts of it. Depending on the DAE structure this approach may lead to a reduction of the index. Here we follow the techniques in [MR99, Est00, Rod00] for DAEs without a properly stated leading term to reduce the index of a subclass of index-2 DAE (2.8) with a properly stated leading term. For index-2 DAEs of the form (2.10) an index reduction result can be found in [Mär01, Men11]. We have already applied these techniques in [Bau08, BST10] for index-2 DAEs of the form (2.8).

The next lemma shows that the obvious constraint set $\mathcal{M}_0\,(t_0)$ describes all constraints in case of an index-1 DAE (2.8).

Theorem 2.33. Let the DAE (2.8) be of index-1 and $t_0 \in \mathcal{I}$. Then through each $y_0 \in \mathcal{M}_0\,(t_0)$ passes exactly one solution to the DAE (2.8).

Proof. See Theorem 2.3 in [HM04], using the relation between the DAE (2.8) and its natural extension given in Theorem 2.16 and 2.29. $\qquad\square$

Note that Theorem 2.33 ensures local unique solvability only. For a a global unique solvability result for DAEs using the concept of strong monotonicity under certain structural conditions we refer to [JMT12].

In contrast to index-1 DAEs the flow of the index-2 DAE (2.8) is additionally restricted by a set $\mathcal{M}_1\,(t)$, where the relation $\mathcal{M}_1\,(t) \subset \mathcal{M}_0\,(t)$ holds true. For every solution $y \in \mathcal{C}\,(\mathcal{I}, \mathbb{R}^n)$ of the DAE (2.8) the relation $y\,(t) \in \mathcal{M}_1\,(t)$, is fulfilled for all $t \in \mathcal{I}$. The set $\mathcal{M}_1\,(t)$ is the *index-2 constraint set* with $\mathcal{M}_1\,(t) = \mathcal{M}_0\,(t) \cap \mathcal{H}_1\,(t)$, $t \in \mathcal{I}$, where $\mathcal{H}_1\,(t)$ is the so-called *hidden constraint set*. In case of an index-2 DAE for a consistent value $y_0 = y\,(t_0) \in \mathcal{M}_1\,(t_0)$ holds true for $(y_0, t_0) \in \mathcal{D} \times \mathcal{I}$.

Example 2.34. Consider the index-2 DAE

$$\frac{d}{dt}u - u = 0 \tag{2.15}$$

$$v\frac{d}{dt}v - vz = 0 \tag{2.16}$$

$$u^2 + v^2 - 1 = 0 \tag{2.17}$$

on $\mathcal{D} = \{(u, v, z) \in \mathbb{R}^3 | v > 0\}$. Obviously we get

$$\mathcal{M}_0\,(t) = \left\{(u, v, z) \in \mathbb{R}^3 | u^2 + v^2 - 1 = 0\right\}.$$

Differentiation of (2.17) and utilizing (2.15) and (2.16) yields the hidden constraint set

$$\mathcal{H}_1\,(t) = \left\{(u, v, z) \in \mathbb{R}^3 | u^2 + vz = 0\right\}.$$

That is, the index-2 constraint set is given by

$$\mathcal{M}_1\,(t) = \left\{(u, v, z) \in \mathbb{R}^3 | u^2 + v^2 - 1 = 0 \text{ and } u^2 + vz = 0\right\}.$$

Next we determine the index-2 constraint set $\mathcal{M}_1(t)$ by the use of index reduction. Let $W_0(y,t)$ and $W_1(z,y,t)$ be projectors along $\operatorname{im} G_0(y,t)$ and $\operatorname{im} G_1(z,y,t)$, respectively. The projectors $W_0(y,t)$ and $W_1(z,y,t)$ will play an important for the index reduction by differentiation. At first we need some basic results presented in the next four lemmata.

Lemma 2.35 (Lemma 2.3.1, [Est00]). Let the DAE (2.8) be given. The identities

(i) $W_1(z,y,t) W_0(y,t) = W_1(z,y,t)$

(ii) $W_1(z,y,t) B_0(z,y,t) Q_0(y,t) = 0$

hold true, for all $(z,y,t) \in \mathbb{R}^m \times \mathcal{D} \times \mathcal{I}$.

Proof. (i) We get

$$0 = W_1(z,y,t) G_1(z,y,t) P_0(y,t) = W_1(z,y,t) G_0(y,t).$$

Hence $\operatorname{im} G_0(y,t) \subset \ker W_1(z,y,t)$ that is

$$\ker W_0(y,t) = \operatorname{im} G_0(y,t) \subset \ker W_1(z,y,t)$$

and we conclude

$$W_1(z,y,t)(I - W_0(y,t)) = 0 \Leftrightarrow W_1(z,y,t) W_0(y,t) = W_1(z,y,t).$$

(ii) We obtain directly

$$\begin{aligned}
0 &= W_1(z,y,t) G_1(z,y,t) \\
&= W_1(z,y,t)(G_0(y,t) + B_0(z,y,t) Q_0(y,t)) \\
&= W_1(z,y,t) B_0(z,y,t) Q_0(y,t).
\end{aligned}$$

$\square$

Lemma 2.36. Let an index-2 DAE (2.8) be given. The identity

$$\operatorname{im} W_1(z,y,t) = \operatorname{im} W_1(z,y,t) B_0(z,y,t)$$

holds true for all $(z,y,t) \in \mathbb{R}^m \times \mathcal{D} \times \mathcal{I}$.

Proof. Clearly $\operatorname{im} W_1(z,y,t) B_0(z,y,t) \subset \operatorname{im} W_1(z,y,t)$ holds true. From the index-2 condition $\mathcal{N}_1(z,y,t) \oplus \mathcal{S}_1(z,y,t) = \mathbb{R}^n$ can be deduced, see [GM86] using a canonical projector

$$Q_{1,S}(z,y,t) = Q_1(z,y,t) G_2(z,y,t)^{-1} B_0(z,y,t) P_0(y,t).$$

Using the Rank–nullity theorem and Lemma 2.35 we can conclude

$$\begin{aligned}
\dim(\operatorname{im} W_1(z,y,t) B_0(z,y,t)) &= \dim(\operatorname{im} W_1(z,y,t) B_0(z,y,t) P_0(y,t)) \\
&= n - \dim(\ker W_1(z,y,t) B_0(z,y,t) P_0(y,t))
\end{aligned}$$

$$\begin{aligned} &= n - \dim \mathcal{S}_1\left(z, y, t\right) \\ &= \dim \mathcal{N}_1\left(z, y, t\right) \\ &= \dim \left(\ker G_1\left(z, y, t\right)\right) \\ &= n - \dim \left(\operatorname{im} G_1\left(z, y, t\right)\right) \\ &= n - \dim \left(\ker W_1\left(z, y, t\right)\right) \\ &= \dim \left(\operatorname{im} W_1\left(z, y, t\right)\right) \end{aligned}$$

and therefore it follows the missing inclusion, see Remark 2.28. □

For the next investigations we denote by $D\left(y, t\right)^-$ a pseudoinverse of $D\left(y, t\right)$. To obtain a unique $D\left(y, t\right)^-$ we choose

$$D\left(y, t\right)^- D\left(y, t\right) = P_0\left(y, t\right) \text{ and } D\left(y, t\right) D\left(y, t\right)^- = R\left(t\right)$$

for all $\left(y, t\right) \in \mathcal{D} \times \mathcal{I}$, see Theorem A.14 and Lemma A.15.

Lemma 2.37. Let an index-2 DAE (2.8) be given. The identity

$$\operatorname{im} W_1\left(z, y, t\right) = \operatorname{im} W_1\left(z, y, t\right) B_0\left(z, y, t\right) D\left(y, t\right)^-$$

holds true for all $\left(z, y, t\right) \in \mathbb{R}^m \times \mathcal{D} \times \mathcal{I}$ with $P_0\left(y, t\right) = D\left(y, t\right)^- D\left(y, t\right)$.

Proof. Clearly one inclusion is obvious. Let be $x \in \operatorname{im} W_1\left(z, y, t\right)$. Then there are $v, u \in \mathbb{R}^n$ and $w \in \mathbb{R}^m$ such that

$$\begin{aligned} x &= W_1\left(z, y, t\right) v \\ &= W_1\left(z, y, t\right) B_0\left(z, y, t\right) P_0\left(y, t\right) u \\ &= W_1\left(z, y, t\right) B_0\left(z, y, t\right) D\left(y, t\right)^- D\left(y, t\right) u \\ &= W_1\left(z, y, t\right) B_0\left(z, y, t\right) D\left(y, t\right)^- w \end{aligned}$$

and hence $x \in \operatorname{im} W_1\left(z, y, t\right) B_0\left(z, y, t\right) D\left(y, t\right)^-$, see Lemma 2.35 and 2.36. □

In the following we need $d\left(y, t\right)$ depends only on dynamic components. This structure will be exploited later on.

Lemma 2.38. Let the DAE (2.8) be given with $P_0 \in \mathcal{C}\left(\mathcal{I}, \mathbb{R}^{n \times n}\right)$ and domain $\mathcal{D}$ so that for each $\left(y, t\right) \in \mathcal{D} \times \mathcal{I}$ also $P_0\left(t\right) y + s Q_0\left(t\right) y \in \mathcal{D}$ for all $s \in [0, 1]$. Then, the identities

(i) $d\left(y, t\right) = d\left(P_0\left(t\right) y, t\right)$

(ii) $D\left(y, t\right) = D\left(P_0\left(t\right) y, t\right)$

(iii) $\frac{\partial}{\partial t} d\left(y, t\right) = D\left(P_0\left(t\right) y, t\right) \frac{d}{dt} P_0\left(t\right) y + \frac{\partial}{\partial t} d\left(P_0\left(t\right) y, t\right)$

hold true for all $\left(y, t\right) \in \mathcal{D} \times \mathcal{I}$.

Proof. We use $\ker P_0(t) = \operatorname{im} Q_0(t) = \ker D(y,t)$ for all $(y,t) \in \mathcal{D} \times \mathcal{I}$. (i) We apply the mean value theorem, see [Mär03]. We get

$$d(y,t) - d(P_0(t)y,t) = \int_0^1 D(sy + (1-s)P_0(t)y,t)Q_0(t)y\,ds = 0.$$

(ii) We directly obtain

$$D(y,t) = \frac{\partial}{\partial y}d(y,t) = \frac{\partial}{\partial y}d(P_0(t)y,t) = D(P_0(t)y,t)P_0(t) = D(P_0(t)y,t).$$

(iii) On the one hand we have

$$\frac{d}{dt}d(y,t) = D(y,t)\frac{d}{dt}y + \frac{\partial}{\partial t}d(y,t)$$

and on the other hand

$$\begin{aligned}
\frac{d}{dt}d(P_0(t)y,t) &= D(P_0(t)y,t)P_0(t)\frac{d}{dt}y + D(P_0(t)y,t)\frac{d}{dt}P_0(t)y + \frac{\partial}{\partial t}d(P_0(t)y,t) \\
&= D(P_0(t)y,t)\frac{d}{dt}y + D(P_0(t)y,t)\frac{d}{dt}P_0(t)y + \frac{\partial}{\partial t}d(P_0(t)y,t).
\end{aligned}$$

Combining both via $\frac{d}{dt}d(y,t) = \frac{d}{dt}d(P_0(t)y,t)$ we achieve

$$\frac{\partial}{\partial t}d(y,t) = D(P_0(t)v,t)\frac{d}{dt}P_0(t)y + \frac{\partial}{\partial t}d(P_0(t)y,t).$$

$\square$

Now we collect all ingredients for the index reduction of index-2 DAEs. The application classes we investigate the forthcoming chapters have special structures. We will restrict ourselves to index-2 DAEs (2.8) of the form

$$A(y,t)\frac{d}{dt}d(y,t) + b(y,t) = 0, \qquad (2.18)$$

where we assume

- constant projectors Q_0 and W_1,

- domain $\mathcal{D}$ so that for each $y \in \mathcal{D}$ also $P_0y + sQ_0y \in \mathcal{D}$ for all $s \in [0,1]$,

- the continuous partial derivative $\frac{\partial}{\partial t}W_1b(y,t)$ exists for all $(y,t) \in \mathcal{D} \times \mathcal{I}$

and

- by $D(y,t)^-$ we denote the pseudoinverse of $D(y,t)$ with $D(y,t)^- D(y,t) = P_0$ and $D(y,t)D(y,t)^- = R(t)$, see above.

24

Remark 2.39. Since W_1 is constant the relation

$$W_1 B_0 (z, y, t) = W_1 \frac{\partial}{\partial y} \left[A (y, t) z + b (y, t) \right] = W_1 \frac{\partial}{\partial y} b (y, t)$$

holds true.

To extract suitable parts of the DAE (2.18) to reduce the index by differentiation we left-multiplying the DAE (2.18) by W_1 and obtain

$$W_1 b (y, t) = 0$$

due to Lemma 2.35 and $\ker W_0 (y, t) = \operatorname{im} G_0 (y, t) = \operatorname{im} A (y, t)$. Hence the relation

$$\frac{d}{dt} W_1 b (y, t) = 0 \tag{2.19}$$

holds true, too. The next step is to describe the derivative (2.19) in a proper way so that the DAE (2.18) can be reformulated as an index-1 DAE with a properly stated leading term.

Lemma 2.40. Let the DAE (2.18) be given. The relation

$$\frac{d}{dt} W_1 b (y, t) = W_1 \frac{\partial}{\partial y} b (y, t) D (y, t)^- \left(\frac{d}{dt} d (y, t) - \frac{\partial}{\partial t} d (y, t) \right) + \frac{\partial}{\partial t} W_1 b (y, t)$$

holds true for all $(y, t) \in \mathcal{D} \times \mathcal{I}$.

Proof. We apply Lemma 2.35 and 2.38. Since Q_0 is constant, it holds that

$$\frac{d}{dt} d (y, t) = D (y, t) \frac{d}{dt} \left[P_0 y \right] + \frac{\partial}{\partial t} d (y, t) \tag{2.20}$$

and

$$\frac{d}{dt} W_1 b (y, t) = W_1 \frac{\partial}{\partial y} b (y, t) \frac{d}{dt} \left[P_0 y \right] + \frac{\partial}{\partial t} W_1 b (y, t) . \tag{2.21}$$

Left-multiplying of (2.20) by $D (y, t)^-$ leads to

$$\frac{d}{dt} \left[P_0 y \right] = D (y, t)^- \frac{d}{dt} d (y, t) - D (y, t)^- \frac{\partial}{\partial t} d (y, t)$$

since $D (y, t)^- D (y, t) = P_0$. Substitution into (2.21) yields the result. $\qquad \square$

The DAE (2.18) can be written as

$$\begin{aligned} 0 &= A (y, t) \frac{d}{dt} d (y, t) + b (y, t) \\ &= A (y, t) \frac{d}{dt} d (y, t) + (I - W_1) b (y, t) + W_1 b (y, t) \end{aligned}$$

and replacing $W_1 b(y,t)$ by $\frac{d}{dt} W_1 b(y,t)$ leads to

$$A(y,t) \frac{d}{dt} d(y,t) + (I - W_1) b(y,t) + \frac{d}{dt} W_1 b(y,t) = 0.$$

Using Lemma 2.40 we obtain from the index-2 DAE (2.18) the index-1 DAE

$$\overline{A}(y,t) \frac{d}{dt} d(y,t) + \overline{b}(y,t) = 0 \tag{2.22}$$

with

$$\overline{A}(y,t) = A(y,t) + W_1 \frac{\partial}{\partial y} b(y,t) D(y,t)^-,$$

$$\overline{b}(y,t) = (I - W_1) b(y,t) - W_1 \frac{\partial}{\partial y} b(y,t) D(y,t)^- \frac{\partial}{\partial t} d(y,t) + \frac{\partial}{\partial t} W_1 b(y,t)$$

and a properly stated leading term. Within the next lemmata we prove the properly stated leading term and that the DAE (2.22) has indeed index-1.

Lemma 2.41. The DAE (2.22) has a properly stated leading term utilizing the projector $R \in \mathcal{C}^1(\mathcal{I}, \mathbb{R}^{n \times n})$ of the DAE (2.18).

Proof. It is sufficient to show the relation $\ker A(y,t) = \ker \overline{A}(y,t)$. The first inclusion $\ker A(y,t) \subset \ker \overline{A}(y,t)$ follows immediately using the identity

$$W_1 \frac{\partial}{\partial y} b(y,t) D(y,t)^- = W_1 \frac{\partial}{\partial y} b(y,t) D(y,t)^- R(t)$$

and $\ker A(y,t) = \ker R(t)$. Let be $v \in \ker \overline{A}(y,t)$. Using the constant projector W_1, see Lemma 2.35, we achieve $v \in \ker W_1 \frac{\partial}{\partial y} b(y,t) D(y,t)^-$ and hence $v \in \ker A(y,t)$. That means the decomposition in $\ker \overline{A}(y,t)$ and $\operatorname{im} D(y,t)$ can be realized using the projector $R(t)$. $\qquad \square$

The relation $\ker G_0(y,t) = \ker \overline{G_0}(y,t)$ holds true since the DAEs (2.18) and (2.22) have the same leading term $d(y,t)$, see Lemma 2.12. That is, the same derivatives occur in both DAEs.

First, we need a technical lemma to handle second order derivatives with respect to the unknowns to prove the index-1 result for the DAE (2.22).

Lemma 2.42. Let the DAE (2.18) be given. Then, the relations

  (i) $W_1 b(y,t) = W_1 b(P_0 y, t)$

  (ii) $\frac{\partial}{\partial y} \left[W_1 \frac{\partial}{\partial y} b(y,t) z \right] Q_0 = 0$ for all $z \in \mathbb{R}^n$

hold true for all $(y,t) \in \mathcal{D} \times \mathcal{I}$.

Proof. (i) We apply the mean value theorem. We get

$$
\begin{aligned}
W_1 b\,(y,t) - W_1 b\,(P_0 y, t) &= \int_0^1 W_1 \frac{\partial}{\partial y} b\,(sy + (1-s)\,P_0 y, t)\,Q_0 y\,ds \\
&= \int_0^1 W_1 B_0\,(z, sy + (1-s)\,P_0 y, t)\,Q_0 y\,ds \\
&= 0.
\end{aligned}
$$

(ii) We define $H\,(y) = W_1 \frac{\partial}{\partial y} b\,(y,t)\,z$ for fixed $z \in \mathbb{R}^n$. Regarding the directional derivative of $H\,(y)$ along $Q_0 w$ for all $w \in \mathbb{R}^n$. Using (i) we get

$$
\begin{aligned}
\frac{\partial}{\partial y} H\,(y)\,Q_0 w &= \lim_{h\to 0} \frac{H\,(y + hQ_0 w) - H\,(y)}{h} \\
&= \lim_{h\to 0} \frac{1}{h}\left[W_1 \frac{\partial}{\partial y} b\,(y + hQ_0 w, t)\,z - W_1 \frac{\partial}{\partial y} b\,(y,t)\,z \right] \\
&= \lim_{h\to 0} \frac{1}{h} \frac{\partial}{\partial y} \left[W_1 b\,(P_0 y, t)\,z - W_1 b\,(P_0 y, t)\,z \right] \\
&= 0
\end{aligned}
$$

for all $w \in \mathbb{R}^n$. $\qquad\qquad\square$

Lemma 2.43. The DAE (2.22) has index-1.

Proof. We compute $\overline{G_1}\,(z,y,t)$ by

$$
\begin{aligned}
\overline{G_1}\,(z,y,t) &= \overline{G_0}\,(y,t) + \overline{B_0}\,(z,y,t)\,Q_0 \\
&= \overline{G_0}\,(y,t) + \frac{\partial}{\partial y}\left[\overline{A}\,(y,t)\,z + \overline{b}\,(y,t) \right] Q_0 \\
&= G_1\,(z,y,t) + W_1 \frac{\partial}{\partial y} b\,(y,t)\,P_0 \\
&\quad + \frac{\partial}{\partial y}\left[W_1 \frac{\partial}{\partial y} b\,(y,t)\,D\,(y,t)^- \left(z - \frac{\partial}{\partial t} d\,(y,t) \right) \right] Q_0
\end{aligned}
$$

deploying Lemma 2.35. Using Lemma 2.42 we get

$$
\overline{G_1}\,(z,y,t) = G_1\,(z,y,t) + W_1 \frac{\partial}{\partial y} b\,(y,t)\,P_0.
$$

Finally we have

$$
\begin{aligned}
v \in \ker \overline{G_1}\,(z,y,t) &\Leftrightarrow v \in \ker G_1\,(z,y,t) \text{ and } v \in \ker W_1 B_0\,(z,y,t)\,P_0 \\
&\Leftrightarrow v \in \mathcal{N}_1\,(z,y,t) \cap \mathcal{S}_1\,(z,y,t)
\end{aligned}
$$

by using Remark 2.39 and hence $v = 0$, because $\mathcal{N}_1\,(z,y,t) \cap \mathcal{S}_1\,(z,y,t) = \{0\}$ due to the DAE (2.18) having index-2. The decomposition of $\overline{G_1}\,(z,y,t)$ can be realized by W_1 and we conclude that $\overline{G_1}\,(z,y,t)$ is nonsingular, that is, the DAE (2.22) has index-1. $\quad\square$

To make use of the index reduced DAE (2.22) we need to relate the solution to (2.18) and (2.22). An analytical solution to the index reduced DAE is not necessarily a solution to the index-2 DAE but the other way holds true. The index reduced DAE has more solutions than the index-2 one. But if the initial conditions are chosen properly a solution to the index reduced DAE is a solution to the index-2 DAE, too.

Theorem 2.44. Let $W_1 b (y_0, t_0) = 0$ be fulfilled in one point $(y_0, t_0) \in \mathcal{D} \times \mathcal{I}$. A solution to the DAE (2.18) through (y_0, t_0) is a solution to the DAE (2.22) through (y_0, t_0) and vice versa.

Proof. Let $y_* \in \mathcal{C}_d^1 (\mathcal{I}, \mathcal{D})$ be a solution to the DAE (2.18). Due to construction the solution is a solution y_* to the DAE (2.22), too.
The other direction is only true if $W_1 b (y_0, t_0) = 0$ for $(y_0, t_0) \in \mathcal{D} \times \mathcal{I}$ and $y (t_0) = y_0$. If the relation $\frac{d}{dt} W_1 b (y, t) = 0$ holds true then $W_1 b (y, t) = 0$ for all $(y, t) \in \mathcal{D} \times \mathcal{I}$. Let $y_* \in \mathcal{C}_d^1 (\mathcal{I}, \mathcal{D})$ be a solution to the DAE (2.22) with $y (t_0) = y_0$. Then

$$\overline{A} (y_*, t) \frac{d}{dt} d (y_*, t) + \overline{b} (y_*, t) = 0$$

and left-multiplication by W_1 yields

$$W_1 \frac{\partial}{\partial y} b (y_*, t) D (y_*, t)^- \left(\frac{d}{dt} d (y_*, t) - \frac{\partial}{\partial t} d (y_*, t) \right) + \frac{\partial}{\partial t} W_1 b (y_*, t) = 0$$

and

$$\frac{d}{dt} W_1 b (y_*, t) = 0$$

respectively, see Lemma 2.40. Due to $W_1 b (y_0, t_0) = 0$ we obtain

$$W_1 b (y_*, t) = 0.$$

Left-multiplying the DAE (2.22) by $(I - W_1)$ results in

$$A (y_*, t) \frac{d}{dt} d (y_*, t) + (I - W_1) b (y_*, t) = 0.$$

Hence y_* is a solution to (2.18). $\qquad\qquad\qquad\qquad\qquad\qquad\qquad\qquad\qquad\quad \square$

Now we are able to describe the hidden constraint set $\mathcal{H}_1 (t)$ and hence the index-2 constraint set $\mathcal{M}_1 (t) = \mathcal{M}_0 (t) \cap \mathcal{H}_1 (t)$ of the DAE (2.18).

Theorem 2.45. The hidden constraint set $\mathcal{H}_1 (t)$ of the DAE (2.18) can be described by

$$\mathcal{H}_1 (t) = \left\{ y \in \mathcal{D} | \exists z \in \mathbb{R}^m : W_1 \frac{\partial}{\partial y} b (y, t) D (y, t)^- \left(z - \frac{\partial}{\partial t} d (y, t) \right) + \frac{\partial}{\partial t} W_1 b (y, t) = 0 \right\}.$$

Proof. The set

$$\overline{\mathcal{M}}_0(t) = \left\{ y(t) \in \mathcal{D} | \overline{b}(y,t) \in \operatorname{im} \overline{A}(y,t) \right\}$$

is the obvious constraint set of the index-1 DAE (2.22). Then the index-2 constraint set $\mathcal{M}_1(t)$ of the DAE (2.18) can be described by

$$\mathcal{M}_1(t) = \left\{ y(t) \in \overline{\mathcal{M}}_0(t) \,|\, W_1 b(y,t) = 0 \right\}.$$

due to Theorem 2.44. Hence

$$\mathcal{M}_1(t) = \left\{ y(t) \in \mathcal{D} | \overline{b}(y,t) \in \operatorname{im} \overline{A}(y,t), W_1 b(y,t) = 0 \right\}$$

$$= \left\{ y(t) \in \mathcal{D} | \exists z \in \mathbb{R}^m : A(y,t) z + (I - W_1) b(y,t), W_1 b(y,t) = 0 \right.$$

$$\left. W_1 \frac{\partial}{\partial y} b(y,t) D(y,t)^- \left(z - \frac{\partial}{\partial t} d(y,t) \right) + \frac{\partial}{\partial t} W_1 b(y,t) = 0 \right\}$$

$$= \mathcal{M}_0(t) \cap \mathcal{H}_1(t)$$

is valid. □

Next we deduce an unique solvability result for index-2 DAEs (2.18) using an unique solvability result for index-1 DAEs (2.8).

Theorem 2.46. Let a DAE (2.18) be given with $t_0 \in \mathcal{I}$. Then through each $y_0 \in \mathcal{M}_1(t_0)$ passes exactly one solution.

Proof. Applying the index reduction by differentiation to the DAE (2.18) we get an index-1 DAE (2.22). Due to Theorem 2.33 for every $y_0 \in \overline{\mathcal{M}}_0(t)$ we obtain a unique solution to the index-1 DAE (2.22). Utilizing Theorem 2.44 leads to the result. □

Remark 2.47. The index-2 constraint set $\mathcal{M}_1(t)$ of the DAE (2.18) is filled with solutions due to the index reduced DAE having index-1 and $\mathcal{M}_1(t) \subset \overline{\mathcal{M}}_0(t)$. Hence every $y \in \mathcal{M}_1(t)$ is a consistent value.

As mentioned previously the index reduced DAE has more solutions than the index-2 DAE. Hence we need to choose proper initial conditions, see Theorem 2.44. Otherwise we provoke the so-called *drift-off-phenomenon*. This is due to differentiating parts of the algebraic constraints of the index-2 DAE. The former algebraic constraints turns into ODEs for the index reduces DAE. That is, the initial values are not restricted by the former algebraic constraints anymore and if the initial values violate the former algebraic constraints then the error increases in time, independent of the step size. Unfortunately consistent initial values are not always available. Additionally the used numerical method do not necessarily preserve the former algebraic constraints even though they are preserved in the ODEs with proper initial values. If the step size goes to zero, the drift-off will go to zero on a fixed time interval, too. To reduce the effect of the drift-off we can apply projection or stabilizing techniques to correct the algebraic constraints at certain time points like a Gear-Gupta-Leimkuhler formulation known for multibody systems, see [Mär96, HNW02, GGL85].

Next we illustrate how the solution to a given example changes due to index reduction.

Example 2.48. Consider the index-2 DAE (2.18) given by

$$\begin{bmatrix} 1 \\ 0 \end{bmatrix} \frac{d}{dt} \left(\begin{bmatrix} 1 & 0 \end{bmatrix} u \right) + \begin{pmatrix} -uv \\ u+1-f(t) \end{pmatrix} = 0$$

with $f(t) \neq 1$ for all $t \in [t_0, T]$, $T \in \mathbb{R}$. The solution is given by

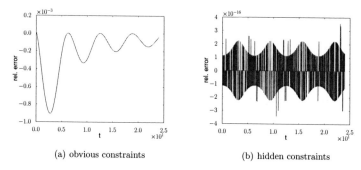

(a) obvious constraints   (b) hidden constraints

Figure 2.2: Index reduced DAE starting with consistent values.

$$u(t) = f(t) - 1 \text{ and } v(t) = \frac{\frac{d}{dt} f(t)}{f(t) - 1}$$

with consistent initial values $u_0 = f(t_0) - 1$ and $v_0 = \frac{\frac{d}{dt} f(t_0)}{f(t_0) - 1}$. The constraints are given

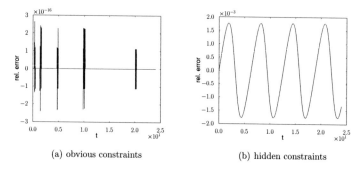

(a) obvious constraints   (b) hidden constraints

Figure 2.3: Index-2 DAE starting with consistent values.

by:

$$u + 1 - f(t) = 0 \text{ (obvious constraint)}$$

$$\text{uv} - \frac{d}{dt} f(t) = 0 \text{ (hidden constraint)}$$

Using the constant projectors

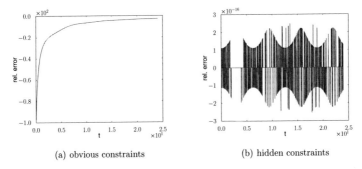

(a) obvious constraints

(b) hidden constraints

Figure 2.4: Index reduced DAE starting with inconsistent values.

$$R = \begin{bmatrix} 1 \end{bmatrix} \text{ and } P_0 = \begin{bmatrix} 1 & 0 \\ 0 & 0 \end{bmatrix}$$

we choose

$$D^- = \begin{bmatrix} 1 \\ 0 \end{bmatrix} \text{ and } W_1 = \begin{bmatrix} 0 & 0 \\ 0 & 1 \end{bmatrix}.$$

This lead to

$$W_1 b \left((u, v), t \right) = \begin{pmatrix} 0 \\ u + 1 - f(t) \end{pmatrix}$$

and

$$\frac{d}{dt} W_1 b \left((u, v), t \right) = \begin{bmatrix} 0 \\ 1 \end{bmatrix} \frac{d}{dt} \left(\begin{bmatrix} 1 & 0 \end{bmatrix} u \right) + \begin{pmatrix} 0 \\ -\frac{d}{dt} f(t) \end{pmatrix}.$$

Hence the index reduced DAE reads

$$\begin{bmatrix} 1 \\ 1 \end{bmatrix} \frac{d}{dt} \left(\begin{bmatrix} 1 & 0 \end{bmatrix} u \right) + \begin{pmatrix} -uv \\ -\frac{d}{dt} f(t) \end{pmatrix} = 0$$

with consistent initial values $u_0 = f(t_0) - 1$ and $v_0 = \frac{\frac{d}{dt} f(t_0)}{f(t_0) - 1}$ fulfilling the hidden constraints. We choose $f(t) = \sin(t) + 3t + 3$, $t_0 = 0$ and $T = 24$. The calculation were carried out by the implicit Euler scheme using the fixed step size $h = 1e\text{-}2$. Starting with consistent initial values the difference of the exact and the numerical solution to the constraints is given in Figure 2.2 and 2.3. To show the drift-off we choose inconsistent initial values $u_0 = -200$ and $v_0 = \frac{\frac{d}{dt} f(t_0)}{f(t_0) - 1}$. The differences in exact and numerical solutions to the constraints are given in Figure 2.4. Note that in case of the index reduced DAE the hidden constraints turn into obvious constraints.

With the index reduction technique we can derive a perturbation result for index-2 DAEs (2.18). First we present a perturbation result for index-1 DAEs (2.8).

Theorem 2.49. Let the DAE (2.8) be of index-1 and $\mathcal{I}_* \subset \mathcal{I}$ a compact interval with $t_0 \in \mathcal{I}_*$. If $y_* \in \mathcal{C}_d^1(\mathcal{I}, \mathcal{D})$ is a solution to the DAE (2.8), then for all solutions $\overline{y} \in \mathcal{C}_d^1(\mathcal{I}, \mathcal{D})$ of

$$A(\overline{y}, t) \frac{d}{dt} d(\overline{y}, t) + b(\overline{y}, t) = q(t),$$

the inequality

$$\|\overline{y} - y_*\|_\infty \leqslant c \left(\|\overline{y}(t_0) - y_*(t_0)\|_\infty + \|q\|_\infty \right)$$

holds true for some $c > 0$ as long as $\|q\|_\infty$ and $\|\overline{y}(t_0) - y_*(t_0)\|_\infty$ are sufficiently small and $q \in \mathcal{C}(\mathcal{I}_*, \mathbb{R}^n)$.

Proof. See Theorem 4.11 and Remark 4.12 in [LMT13] and using the relation between the DAE (2.8) and its natural extension given in Theorem 2.16 and 2.29. $\square$

That is, if the DAE (2.8) has index-1, then the DAE (2.8) has perturbation index-1, see Definition 2.10. With that preliminary work we elaborate a new perturbation result for index-2 DAEs (2.18) using the index reduction techniques.

Theorem 2.50. Let the DAE (2.18) be of index-2 and $\mathcal{I}_* \subset \mathcal{I}$ a compact interval with $t_0 \in \mathcal{I}_*$. If $y_* \in \mathcal{C}_d^1(\mathcal{I}, \mathcal{D})$ is a solution to the DAE (2.18), then for all solutions $\overline{y} \in \mathcal{C}_d^1(\mathcal{I}, \mathcal{D})$ of

$$A(\overline{y}, t) \frac{d}{dt} d(\overline{y}, t) + b(\overline{y}, t) = q(t), \tag{2.23}$$

the inequality

$$\|\overline{y} - y_*\|_\infty \leqslant c \left(\|\overline{y}(t_0) - y_*(t_0)\|_\infty + \|q\|_\infty + \left\| \frac{d}{dt} q \right\|_\infty \right)$$

holds true for some $c > 0$ as long as $\|q\|_\infty$, $\left\| \frac{d}{dt} q \right\|_\infty$ and $\|\overline{y}(t_0) - y_*(t_0)\|_\infty$ are sufficiently small and $q \in \mathcal{C}^1(\mathcal{I}_*, \mathbb{R}^n)$.

Proof. Applying the index reduction by differentiation to the DAE (2.18) we get an index-1 DAE (2.22) with

$$\overline{A}(y, t) \frac{d}{dt} d(y, t) + \overline{b}(y, t) = 0 \tag{2.24}$$

and from the perturbed DAE (2.23) we obtain an index-1 DAE (2.22) given by

$$\overline{A}\,(\overline{y}, t)\,\frac{d}{dt}d\,(\overline{y}, t) + \overline{b}\,(\overline{y}, t) = \overline{q}t \qquad (2.25)$$

with $\overline{q} = (I - W_1)\,q + W_1\frac{d}{dt}q$. Assume $\|q\|_\infty$, $\left\|\frac{d}{dt}q\right\|_\infty$ and $\|\overline{y}\,(t_0) - y_*\,(t_0)\|_\infty$ are sufficiently small and $q \in C^1\,(\mathcal{I}_*, \mathbb{R}^n)$. Next we apply Theorem 2.49 and we get the the inequality

$$\|\overline{y} - y_*\|_\infty \leqslant c\,(\|\overline{y}\,(t_0) - y_*\,(t_0)\|_\infty + \|\overline{q}\|_\infty)$$

with $c > 0$, where $y_* \in C^1_d\,(\mathcal{I}, \mathcal{D})$ is a solution to (2.24) and $\overline{y} \in C^1_d\,(\mathcal{I}, \mathcal{D})$ of (2.25). Thus the inequality

$$\|\overline{y} - y_*\|_\infty \leqslant d\left(\|\overline{y}\,(t_0) - y_*\,(t_0)\|_\infty + \|q\|_\infty + \left\|\frac{d}{dt}q\right\|_\infty\right)$$

with $d > 0$ holds true, due to the projector W_1 being constant. $\qquad\square$

That is, if the DAE (2.18) has index-2 then the DAE (2.18) has perturbation index-2, too. That is a major justification for choosing the tractability index as index concept.

2.2.2 Consistent Initialization

An important task for a successful time integration of DAEs is the determination of a consistent initial value. In this subsection we present methods for the calculation of consistent initial values for a subclass of index-1 and index-2 DAEs. For the index-2 case we follow the idea of [Est00]. In contrast to [Est00] we elaborate the approach for DAEs with a properly stated leading term.

For the index-1 case a very general approach to calculate consistent initial values can be given, see [Mär03].

Theorem 2.51. Let the index-1 DAE (2.8) be given and $t_0 \in \mathcal{I}$. The system

$$A(y_0, t_0) z_0 + b(y_0, t_0) = 0 \qquad (2.26)$$

$$(I - R(t_0)) z_0 + R(t_0) \left(d(y_0, t_0) - z^0 \right) = 0 \qquad (2.27)$$

is locally uniquely solvable for given $z^0 \in \mathbb{R}^m$ and provides a consistent initialization $(z_0, y_0, t_0) \in \operatorname{im} R(t_0) \times \mathcal{M}_0(t_0) \times \mathcal{I}$.

Proof. The Jacobian of the nonlinear system (2.26) and (2.26) reads

$$J(z, y) = \begin{bmatrix} A(y, t_0) & B_0(z, y, t_0) \\ I - R(t_0) & R(t_0) D(y, t_0) \end{bmatrix}.$$

Let be $(z_*, y_*) \in \ker J(z, y)$. Then

$$A(y, t_0) z_* + B_0(z, y, t_0) y_* = 0 \qquad (2.28)$$

$$(I - R(t_0)) z_* + R(t_0) D(y, t_0) y_* = 0 \qquad (2.29)$$

and we can conclude that $y_* \in \mathcal{N}_0(y, t) \cap \mathcal{S}_0(z, y, t)$. That is true since (2.28) leads to $B_0(z, y, t_0) y_* \in \operatorname{im} A(y, t_0)$ and (2.29) results in $y_* \in \ker D(y, t_0)$. We achieve $y_* = 0$ because the DAE has index-1. Next (2.28) and (2.29) come to $z_* \in \operatorname{im} R(t_0) \cap \ker R(t_0)$. Hence the Jacobian is nonsingular. From (2.26) we obtain $z_0 \in \operatorname{im} R(t_0)$. $\qquad \square$

In Theorem 2.51 the equation (2.26) ensures that the DAE (2.8) including the obvious constraints are fulfilled and (2.27) provide the uniqueness of z_0.

A difficulty for index-2 DAEs is the description of the so-called *index-2 components*, which belong to the index-1 set $\mathcal{N}_0(y, t) \cap \mathcal{S}_0(z, y, t)$ and are determined neither by differential nor by derivate-free equations but require inherent differentiation. We call the components of the index-1 set index-2 components since the index-1 set would be empty if the DAE may of index-1. To describe the index-2 components we introduce the projector $T(z, y, t)$ onto $\mathcal{N}_0(y, t) \cap \mathcal{S}_0(z, y, t)$ and the complementary projector $U(z, y, t) = I - T(z, y, t)$, see [Tis96].

Example 2.52. To clear clear up the misunderstanding that every single solution component belongs to exactly one index set we inspect the linear index-2 DAE

$$\frac{d}{dt} x_1 - x_2 - x_3 = 0$$

$$x_1 = f(t)$$

$$x_2 - x_3 = 0$$

proposed by [Jan12b]. For the description of the index-2 components we make use of the relation $\operatorname{im} Q_0 Q_1 = \mathcal{N}_0 \cap \mathcal{S}_0$, see Lemma 3.5 in [Tis96]. Choosing the projectors

$$Q_0 = \begin{bmatrix} 0 & 0 & 0 \\ 0 & 1 & 0 \\ 0 & 0 & 1 \end{bmatrix} \text{ and } Q_1 = \begin{bmatrix} 1 & 0 & 0 \\ \frac{1}{2} & 0 & 0 \\ \frac{1}{2} & 0 & 0 \end{bmatrix}$$

we obtain

$$\mathcal{N}_0 \cap \mathcal{S}_0 = \operatorname{span}\left((0,1,1)\right) = \operatorname{im} \mathrm{T} \text{ with } \mathrm{T} = \begin{bmatrix} 0 & 0 & 0 \\ 0 & 1 & 0 \\ 0 & 1 & 0 \end{bmatrix},$$

that is, the index-2 component is a linear combination of the solution components.

The first relation of the next lemma is taken from [Voi06] and is used to exploit the index-2 components.

Lemma 2.53. Let a DAE (2.8) be given. The projector $\mathrm{U}\left(z,y,t\right)$ can be chosen so that

(i) $\mathrm{U}\left(z,y,t\right)\mathrm{P}_0\left(y,t\right) = \mathrm{P}_0\left(y,t\right) = \mathrm{P}_0\left(y,t\right)\mathrm{U}\left(z,y,t\right)$

(ii) $\mathrm{P}_0\left(y,t\right)\mathrm{P}_1\left(z,y,t\right)\mathrm{U}\left(z,y,t\right) = \mathrm{P}_0\left(y,t\right)\mathrm{P}_1\left(z,y,t\right)$

hold true for all $(z,y,t) \in \mathbb{R}^m \times \mathcal{D} \times \mathcal{I}$.

Proof. (i) With $\operatorname{im}\mathrm{T}\left(z,y,t\right) \subset \ker\mathrm{P}_0\left(y,t\right)$ we get $\mathrm{P}_0\left(y,t\right)\mathrm{T}\left(z,y,t\right) = 0$. Furthermore we can choose the projector $\mathrm{T}\left(z,y,t\right)$ with the property $\mathrm{T}\left(z,y,t\right)\mathrm{P}_0\left(y,t\right) = 0$ due to $\operatorname{im}\mathrm{P}_0\left(y,t\right) \cap \operatorname{im}\mathrm{T}\left(z,y,t\right) = \{0\}$.

(ii) Using (i) we get

$$\mathrm{P}_0\left(y,t\right)\mathrm{P}_1\left(z,y,t\right)\mathrm{P}_0\left(y,t\right)\mathrm{U}\left(z,y,t\right) = \mathrm{P}_0\left(y,t\right)\mathrm{P}_1\left(z,y,t\right)\mathrm{P}_0\left(y,t\right).$$

Furthermore we have

$$\begin{aligned} \mathrm{P}_0\left(y,t\right)\mathrm{P}_1\left(z,y,t\right)\mathrm{P}_0\left(y,t\right) &= \left(\mathrm{I} - \mathrm{Q}_0\left(y,t\right)\right)\left(\mathrm{I} - \mathrm{Q}_1\left(z,y,t\right)\right)\left(\mathrm{I} - \mathrm{Q}_0\left(y,t\right)\right) \\ &= \left(\mathrm{I} - \mathrm{Q}_0\left(y,t\right)\right)\left(\mathrm{I} - \mathrm{Q}_1\left(z,y,t\right)\right) \\ &= \mathrm{P}_0\left(y,t\right)\mathrm{P}_1\left(z,y,t\right) \end{aligned}$$

due to $\mathrm{Q}_1\left(z,y,t\right)\mathrm{Q}_0\left(y,t\right) = 0$. □

We restrict ourselves to index-2 DAEs (2.18) of the form

$$\mathrm{A}\left(t\right)\frac{\mathrm{d}}{\mathrm{d}t}\mathrm{d}\left(y,t\right) + \mathrm{b}\left(y,t\right) = 0, \tag{2.30}$$

where we assume

- $\mathcal{N}_0 \cap \mathcal{S}_0\left(y,t\right)$ does not depend on $(y,t) \in \mathcal{D} \times \mathcal{I}$ and T is a constant projector onto the index-1 set, $\mathrm{U} = \mathrm{I} - \mathrm{T}$ and $\mathrm{P}_0 = \mathrm{UP}_0$, see [Est00],

- domain $\mathcal{D}$ so that for each $y \in \mathcal{D}$ also $\mathrm{U}y + s\mathrm{T}y \in \mathcal{D}$ for all $s \in [0,1]$,

and

- index-2 components Ty occur linearly only, that is, the DAE (2.30) can be written as

$$A(t) \frac{d}{dt} d(Uy, t) + \overline{b}(Uy, t) + B(t) Ty = 0,$$

where $b(y, t) = \overline{b}(Uy, t) + B(t) Ty$ and $B(t) \in \mathcal{C}(\mathcal{I}, \mathbb{R}^{n \times n})$ problem given.

In this subsection we develop a step-by-step method to compute consistent initial values guided by [Est00]. Note that we extend the approach to DAEs with a properly stated leading term. Under the later structural properties we compute a consistent initialization for index-2 DAEs as follows:

(i) Describe the hidden constraints.

(ii) Compute an operating point.

(iii) Correct the operating point to fulfill the hidden constraints.

Next we derive the necessary statements for this step-by-step approach. The leading term $d(y, t)$ of the DAE (2.8) depends only on the non-index-2 components as shown in the following which is an essential ingredient for our investigations. The next two lemmata are used to describe the hidden constraint set $\mathcal{H}_1(t)$ given in Theorem 2.45 without the index-2 components.

Lemma 2.54. Let the DAE (2.8) be given with $P_0, U \in \mathcal{C}(\mathcal{I}, \mathbb{R}^{n \times n})$ and domain $\mathcal{D}$ so that for each $(y, t) \in \mathcal{D} \times \mathcal{I}$ also $P_0(t) y + s Q_0(t) y \in \mathcal{D}$ and $U(t) y + s T(t) y \in \mathcal{D}$ for all $s \in [0, 1]$. The identities

(i) $d(y, t) = d(U(t) y, t)$

(ii) $D(y, t) = D(U(t) y, t)$

(iii) $D(y, t)^- = D(U(t) y, t)^-$

(iv) $\frac{\partial}{\partial t} d(y, t) = D(U(t) y, t) \frac{d}{dt} U(t) y + \frac{\partial}{\partial t} d(U(t) y, t)$

hold true for all $(y, t) \in \mathcal{D} \times \mathcal{I}$.

Proof. Following the proof of Lemma 2.38 and using Lemma 2.53 leads to (i), (ii) and (iv). (iii) Using (ii) the matrix $D(y, t)^-$ fulfills the conditions to be a pseudoinverse of $D(U(t) y, t)$. □

Lemma 2.55 (Lemma 2.3.4, [Est00]). Let the DAE (2.8) be given with domain $\mathcal{D}$ so that for each $(y, t) \in \mathcal{D} \times \mathcal{I}$ also $U(t) y + s T(t) y \in \mathcal{D}$ for all $s \in [0, 1]$, $W_0, U \in \mathcal{C}(\mathcal{I}, \mathbb{R}^{n \times n})$ and $W_1 \in \mathbb{R}^{n \times n}$. Then

(i) $W_0(t) B_0(y, t) T(t) = 0$

(ii) $W_0(t) b(y, t) = W_0(t) b(U(t) y, t)$

(iii) $W_0(t) B_0(y, t) = W_0(t) B_0(U(t) y, t) U(t)$

(iv) $\frac{\partial}{\partial t} W_1 b(y, t) = \frac{\partial}{\partial t} W_1 b(U(t) y, t) + W_1 B_0(U(t) y, t) \frac{d}{dt} U(t) y$

hold true for all $(y, t) \in \mathcal{D} \times \mathcal{I}$.

Proof. (i) We have $\operatorname{im} T(t) \subset \mathcal{S}_0(y, t)$ and $\ker W_0(t) B_0(y, t) = \mathcal{S}_0(y, t)$, see Remark 2.27, that is, $W_0(t) B_0(y, t) T(t) = 0$.
(ii) We apply the mean value theorem. We get

$$W_0(t) b(y, t) - W_0(t) b(Uy, t) = \int_0^1 W_0(t) B_0(sy + (1-s) U(t) y, t) T(t) y \, ds = 0$$

since (i) holds.
(iii) We have

$$\begin{aligned}
W_0(t) B_0(y, t) &= \frac{\partial}{\partial y} W_0(t) b(y, t) \\
&= \frac{\partial}{\partial y} W_0(t) b(U(t) y, t) \\
&= W_0(t) B_0(U(t) y, t) U(t).
\end{aligned}$$

(iv) On the one hand we have

$$\frac{d}{dt} W_1 b(y, t) = W_1 B_0(y, t) \frac{d}{dt} y + \frac{\partial}{\partial t} W_1 b(y, t)$$

and on the other hand applying (iii) we get

$$\begin{aligned}
\frac{d}{dt} W_1 b(U(t) y, t) &= W_1 B_0(U(t) y, t) U(t) \frac{d}{dt} y + W_1 B_0(U(t) y, t) \frac{d}{dt} U(t) y \\
&\quad + \frac{\partial}{\partial t} W_1 b(U(t) y, t) \\
&= W_1 B_0(U(t) y, t) \frac{d}{dt} y + W_1 B_0(U(t) y, t) \frac{d}{dt} U(t) y \\
&\quad + \frac{\partial}{\partial t} W_1 b(U(t) y, t).
\end{aligned}$$

Combining both via $\frac{d}{dt} W_1 b(y, t) = \frac{d}{dt} W_1 b(U(t) y, t)$ proves the statement. $\square$

We have already described the hidden constraints for the DAE (2.30), see Theorem 2.45, but have not yet taken into account the projector U.

Theorem 2.56. The hidden constraint set $\mathcal{H}_1(t)$ of the DAE (2.30) can be described as

$$\begin{aligned}
\mathcal{H}_1(t) = \Big\{ y \in \mathcal{D} \,\big|\, \exists z \in \mathbb{R}^m : W_1 B_0(Uy, t) D(Uy, t)^- \Big(z - \frac{\partial}{\partial t} d(Uy, t) \Big) \\
+ \frac{\partial}{\partial t} W_1 b(Uy, t) = 0 \Big\}.
\end{aligned}$$

Proof. Let be $y \in \mathcal{H}_1 (t)$. Then there is $z \in \mathbb{R}^m$ with

$$0 = W_1 B_0 (y, t) D (y, t)^- \left(z - \frac{\partial}{\partial t} d (y, t) \right) + \frac{\partial}{\partial t} W_1 b (y, t),$$

see Theorem 2.45. The use of U being constant, $W_1 = W_1 W_0 (t)$, see Lemma 2.35,

$$UD (Uy, t)^- = UP_0 D (Uy, t)^- = P_0 D (Uy, t)^- = D (Uy, t)^-,$$

due to Lemma 2.53 and $P_0 = D (Uy, t)^- D (Uy, t)$, Lemma 2.54 and 2.55 lead to

$$0 = W_1 B_0 (y, t) D (y, t)^- \left(z - \frac{\partial}{\partial t} d (y, t) \right) + \frac{\partial}{\partial t} W_1 b (y, t)$$
$$= W_1 B_0 (Uy, t) D (Uy, t)^- \left(z - \frac{\partial}{\partial t} d (Uy, t) \right) + \frac{\partial}{\partial t} W_1 b (Uy, t).$$

$\square$

Lemma 2.57. Let the DAE (2.30) be given with $b (y, t) = \overline{b} (Uy, t) + B (t) Ty$. Then

$$B_0 (y, t) T = B (t) T$$

holds true for all $(y, t) \in \mathcal{D} \times \mathcal{I}$.

Proof. We get

$$B_0 (y, t) = \frac{\partial}{\partial y} \left[\overline{b} (Uy, t) + B (t) Ty \right]$$
$$= \frac{\partial}{\partial Uy} \overline{b} (Uy, t) U + B (t) T.$$

Thus right-multiplying by T yields $B_0 (y, t) T = B (t) T$, since $UT = 0$. $\square$

We are ready for the calculation of a consistent initialization in case of an index-2 DAE (2.30). For this we specify and fix all non-index-2 components Uy and try to correct the index-2 components Ty of the DAE (2.30). At first we need an operating point (z^0, y^0, t_0), that is, a triple fulfilling

$$A (t_0) z^0 + \overline{b} \left(Uy^0, t_0 \right) + B (t_0) Ty^0 = 0, \tag{2.31}$$

see Definition 2.18. A consistent initialization (z_0, y_0, t_0), see Definition 2.20, needs to fulfill all constraints and we obtain

$$A (t_0) z_0 + \overline{b} (Uy_0, t_0) + B (t_0) Ty_0 = 0. \tag{2.32}$$

Due to the fixing of the non index-2 components we have

$$Uy^0 = Uy_0 \tag{2.33}$$

and subtraction of (2.31) from (2.32) using (2.33) yields

$$A\left(t_0\right)\left(z_0 - z^0\right) + B\left(t_0\right) T\left(y_0 - y^0\right) = 0. \tag{2.34}$$

In addition the hidden constraints

$$W_1 B_0\left(U y_0, t\right) D\left(U y_0, t_0\right)^-\left(z_0 - \frac{\partial}{\partial t} d\left(U y_0, t_0\right)\right) + \frac{\partial}{\partial t} W_1 b\left(U y_0, t_0\right) = 0 \tag{2.35}$$

are fulfilled, see Theorem 2.56. Using the properties of W_1, Lemma 2.35, 2.55, 2.57 and (2.33) the two equations (2.34) and (2.35) are equivalent to

$$\left(A\left(t_0\right) + W_1 B_0\left(U y^0, t_0\right) D\left(U y^0, t_0\right)^-\right) z + B\left(t_0\right) T y =$$
$$-W_1 B_0\left(U y^0, t_0\right) D\left(U y^0, t_0\right)^-\left(z^0 - \frac{\partial}{\partial t} d\left(U y^0, t_0\right)\right) - \frac{\partial}{\partial t} W_1 b\left(U y^0, t_0\right)$$

with $z = z_0 - z^0$ and $y = y_0 - y^0$.

Now we are able to calculate a consistent initialization starting from an operating point.

Theorem 2.58. Let (z^0, y^0, t_0) be an operating point of the index-2 DAE (2.30). The rectangular linear system

$$\left(A\left(t_0\right) + W_1 B_0\left(U y^0, t_0\right) D\left(U y^0, t_0\right)^-\right) z + B\left(t_0\right) T y =$$
$$-W_1 B_0\left(U y^0, t_0\right) D\left(U y^0, t_0\right)^-\left(z^0 - \frac{\partial}{\partial t} d\left(U y^0, t_0\right)\right) - \frac{\partial}{\partial t} W_1 b\left(U y^0, t_0\right)$$
$$U y = 0$$
$$\left(I - R\left(t_0\right)\right) z = -\left(I - R\left(t_0\right)\right) z^0$$

has a unique solution $(z, y) \in \mathbb{R}^{m+n}$. A consistent initialization (z_0, y_0, t_0) is given by $z_0 = z + z^0$ and $y_0 = y + y^0$.

Proof. The proof is divided into three part: Show that the problem has at most one solution, show that the problem has at least one solution and prove that (z_0, y_0, t_0) is a consistent initialization.

First we show that the matrix

$$M = \begin{bmatrix} A\left(t_0\right) + W_1 B_0\left(U y^0, t_0\right) D\left(U y^0, t_0\right)^- & B\left(t_0\right) T \\ 0 & U \\ \left(I - R\left(t_0\right)\right) & 0 \end{bmatrix} \tag{2.36}$$

is injective, that is,

$$\left(A\left(t_0\right) + W_1 B_0\left(U y^0, t_0\right) D\left(U y^0, t_0\right)^-\right) z + B\left(t_0\right) T y = 0 \tag{2.37}$$

$$Uy = 0 \qquad (2.38)$$
$$(I - R(t_0))z = 0 \qquad (2.39)$$

has only the trivial solution. We split (2.37) into

$$A(t_0)z + B(t_0)Ty = 0 \qquad (2.40)$$
$$W_1 B_0 \left(Uy^0, t_0 \right) D \left(Uy^0, t_0 \right)^{-} z = 0 \qquad (2.41)$$

using W_1, Lemma 2.35 and 2.55. First, we focus on (2.40), which can be rewritten as

$$
\begin{aligned}
0 &= A(t_0)z + B(t_0)Ty \\
&= A(t_0)z + B_0 \left(y^0, t_0 \right) Ty \\
&= A(t_0)R(t_0)z + B_0 \left(y^0, t_0 \right) Ty \\
&= A(t_0)D \left(y^0, t_0 \right) D \left(y^0, t_0 \right)^{-} z + B_0 \left(y^0, t_0 \right) Ty \\
&= G_0 \left(y^0, t_0 \right) D \left(y^0, t_0 \right)^{-} z + B_0 \left(y^0, t_0 \right) Ty
\end{aligned}
$$

by Lemma 2.57. Left multiplying by $G_2 \left(y^0, t_0 \right)^{-1}$ yields

$$
\begin{aligned}
0 &= G_2 \left(y^0, t_0 \right)^{-1} \left(G_0 \left(y^0, t_0 \right) D \left(y^0, t_0 \right)^{-} z + B_0 \left(y^0, t_0 \right) Ty \right) \\
&= P_1 \left(y^0, t_0 \right) P_0 D \left(y^0, t_0 \right)^{-} z + G_2 \left(y^0, t_0 \right)^{-1} B_0 \left(y^0, t_0 \right) P_0 P_1 \left(y^0, t_0 \right) Ty \\
&\quad + Q_1 \left(y^0, t_0 \right) Ty + Q_0 Ty,
\end{aligned}
$$

see Lemma 2.32. Finally, we obtain

$$0 = P_1 \left(y^0, t_0 \right) D \left(y^0, t_0 \right)^{-} z + y, \qquad (2.42)$$

since Lemma 2.53 and (2.38) lead to

$$
\begin{aligned}
Q_0 T &= T \\
Ty &= y \\
Q_1 \left(y^0, t_0 \right) T &= 0 \\
P_0 P_1 \left(y^0, t_0 \right) T &= 0
\end{aligned}
$$

due to $Q_1 \left(y^0, t_0 \right) Q_0 = 0$. Splitting (2.42) by Q_0 and P_0 results in

$$0 = P_0 P_1 \left(y^0, t_0 \right) D \left(y^0, t_0 \right)^{-} z \Leftrightarrow 0 = D \left(y^0, t_0 \right) P_1 \left(y^0, t_0 \right) D \left(y^0, t_0 \right)^{-} z \qquad (2.43)$$

and

$$y = -Q_0 P_1 \left(y^0, t_0 \right) D \left(y^0, t_0 \right)^{-} z \Leftrightarrow y = Q_0 Q_1 \left(y^0, t_0 \right) D \left(y^0, t_0 \right)^{-} z \qquad (2.44)$$

since $Q_0 D \left(y^0, t_0 \right)^{-} = 0$ and $P_0 = D \left(y^0, t_0 \right)^{-} D \left(y^0, t_0 \right)$. By combining (2.39) and (2.43) we gain

$$z = D \left(y^0, t_0 \right) Q_1 \left(y^0, t_0 \right) D \left(y^0, t_0 \right)^{-} z \qquad (2.45)$$

because of

$$z = R(t_0) z$$
$$= D(y^0, t_0) D(y^0, t_0)^- z$$
$$= D(y^0, t_0) P_1(y^0, t_0) D(y^0, t_0)^- z + D(y^0, t_0) Q_1(y^0, t_0) D(y^0, t_0)^- z.$$

Starting from (2.41) using Lemma 2.53, 2.55 and (2.45) we can deduce

$$0 = W_1 B_0(Uy^0, t_0) D(Uy^0, t_0)^- z$$
$$= W_1 B_0(Uy^0, t_0) D(Uy^0, t_0)^- D(y^0, t_0) Q_1(y^0, t_0) D(y^0, t_0)^- z$$
$$= W_1 B_0(Uy^0, t_0) P_0 Q_1(y^0, t_0) D(y^0, t_0)^- z$$
$$= W_1 B_0(y^0, t_0) P_0 Q_1(y^0, t_0) D(y^0, t_0)^- z,$$

that is,

$$Q_1(y^0, t_0) D(y^0, t_0)^- z \in \ker W_1 B_0(y^0, t_0) P_0 = \mathcal{S}_1(y^0, t_0).$$

Furthermore

$$Q_1(y^0, t_0) D(y^0, t_0)^- z \in \operatorname{im} Q_1(y^0, t_0) = \mathcal{N}_1(y^0, t_0)$$

and we conclude $Q_1(y^0, t_0) D(y^0, t_0)^- z = 0$ due to $\mathcal{N}_1(y^0, t_0) \cap \mathcal{S}_1(y^0, t_0) = \{0\}$ since the DAE (2.30) has index-2. From (2.44) and (2.45) we end up with $(z, y) = 0$. That is, the matrix (2.36) is injective and if a solution to the linear system exists then the solution is unique.

The right-hand side of the linear system reads

$$g = \begin{pmatrix} -W_1 B_0(Uy^0, t) D(Uy^0, t_0)^- \left(z^0 - \frac{\partial}{\partial t} d(Uy^0, t_0)\right) - \frac{\partial}{\partial t} W_1 b(Uy^0, t_0) \\ 0 \\ -(I - R(t_0)) z^0 \end{pmatrix}.$$

For the existence of a solution we have to prove $g \in \operatorname{im} M$, that is, we have to show that it exist $(z, y) \in \mathbb{R}^{m+n}$ with

$$\begin{bmatrix} A(t_0) + W_1 B_0(Uy^0, t_0) D(Uy^0, t_0)^- & B(t_0) T \\ 0 & U \\ (I - R(t_0)) & 0 \end{bmatrix} \begin{pmatrix} z \\ y \end{pmatrix} = g.$$

We need $y \in \ker U$ and $z = v - z^0$ with $v \in \operatorname{im} R(t_0)$ and $z^0 \in \mathbb{R}^m$. Using W_1 we split the first equation of the linear system into

$$A(t_0) z + B(t_0) T y = 0 \tag{2.46}$$
$$W_1 B_0(Uy^0, t_0) D(Uy^0, t_0)^- z = -W_1 B_0(Uy^0, t) D(Uy^0, t_0)^- [z^0$$

$$-\frac{\partial}{\partial t} d\left(Uy^0, t_0\right)\Big] - \frac{\partial}{\partial t} W_1 b\left(Uy^0, t_0\right)$$

$$= -W_1 B_0\left(Uy^0, t\right) D\left(Uy^0, t_0\right)^- \left[R\left(t_0\right) z^0 \qquad (2.47)\right.$$

$$\left. - R\left(t_0\right) \frac{\partial}{\partial t} d\left(Uy^0, t_0\right) + R\left(t_0\right) w\right]$$

since $D\left(Uy^0, t_0\right)^- R\left(t_0\right) = D\left(Uy^0, t_0\right)^-$ and it exists $w \in \mathbb{R}^m$ such that

$$W_1 B_0\left(Uy^0, t\right) D\left(Uy^0, t_0\right)^- w = -\frac{\partial}{\partial t} W_1 b\left(Uy^0, t_0\right),$$

see Lemma 2.37. From (2.47) we can deduce

$$v = R\left(t_0\right) \frac{\partial}{\partial t} d\left(Uy^0, t_0\right) - R\left(t_0\right) w \in \operatorname{im} R\left(t_0\right)$$

is always a valid choice for fixing the component z. It remains to fix y. For that we have to investigate (2.46). We get

$$A\left(t_0\right) z + B\left(t_0\right) Ty = A\left(t_0\right) R\left(t_0\right) z + B\left(t_0\right) Ty$$
$$= A\left(t_0\right) R\left(t_0\right) z + B_0\left(y^0, t_0\right) Ty$$
$$= G_0\left(y^0, t_0\right) D\left(y^0, t_0\right)^- z + B_0\left(y^0, t_0\right) Ty$$

and left-multiplication of $G_2^{-1}\left(y^0, t_0\right)$ yields

$$y = -TP_1\left(y^0, t_0\right) D\left(y^0, t_0\right)^- z \in \operatorname{im} T,$$

see above. We show that $g \in \operatorname{im} M$ and hence the system has a unique solution.

In the final step we have to show that $z_0 = z + z^0$ and $y_0 = y + y^0$ is a consistent initialization. Since (z^0, y^0, t_0) is an operating point

$$A\left(t_0\right) z^0 + \overline{b}\left(Uy^0, t_0\right) + B\left(t_0\right) Ty^0 = 0$$

is valid. Addition of $A\left(t_0\right) z + B\left(t_0\right) Ty = 0$ leads to

$$A\left(t_0\right) z_0 + \overline{b}\left(Uy_0, t\right) + B\left(t_0\right) Ty_0 = 0$$

due to $Uy = 0$, that is, (z_0, y_0, t_0) is an operating point and we get $y_0 \in \mathcal{M}_0\left(t_0\right)$. The equation

$$0 = W_1 B_0\left(Uy^0, t_0\right) D\left(Uy^0, t_0\right)^- \left(\left(z + z^0\right) - \frac{\partial}{\partial t} d\left(Uy^0, t_0\right)\right) + \frac{\partial}{\partial t} W_1 b\left(Uy^0, t_0\right)$$

$$= W_1 B_0\left(Uy_0, t_0\right) D\left(Uy_0, t_0\right)^- \left(z_0 - \frac{\partial}{\partial t} d\left(Uy_0, t_0\right)\right) + \frac{\partial}{\partial t} W_1 b\left(Uy_0, t_0\right)$$

ensures that the hidden constraints are fulfilled, that is, $y_0 \in \mathcal{M}_1\left(t_0\right)$ and (z_0, y_0, t_0) is a consistent initialization since $z_0 \in \operatorname{im} R\left(t_0\right)$. $\qquad \square$

The rectangular linear system stated in Theorem 2.58 is well suited for a least square method due to the full column rank.

So far we have not shown how to calculate an operating point (z^0, y^0, t_0), which is an essential ingredient for Theorem 2.58. Motivated by our application classes we determine an operating point for a subclass of DAEs (2.8) given by

$$A(y, t) \frac{d}{dt} d(y, t) + b(y, t) = 0, \qquad (2.48)$$

where we assume

- $\ker B_0(y, t) = \ker B_0(y, t)^\top$ independent of $y \in \mathbb{R}^n$

and

- it exists $y^0 \in \mathbb{R}^n$ such that $b(y^0, t_0) = 0$ for $t_0 \in \mathcal{I}$.

First, we are interested in a *point of equilibrium* $(y^0, t_0) \in \mathbb{R}^n \times \mathcal{I}$ of the DAE (2.48), that is, a point fulfilling $b(y^0, t_0) = 0$. If $y^0 \in \mathcal{D}$, then the point of equilibrium is an operating point (z^0, y^0, t_0) of the DAE (2.48) with $z^0 = 0$. For the calculation of a point of equilibrium for the subclass of DAEs (2.48) we make use of an orthogonal projector decomposition developed in [Jan12a] for a new index concept.

Theorem 2.59. Let

$$f(y, t) = 0 \text{ with } f : \mathbb{R}^n \times \mathbb{R} \to \mathbb{R}^n$$

be given. Moreover, let $F(y, t) = \frac{\partial}{\partial y} f(y, t)$ with $\ker F(y, t) = \ker F(y, t)^\top$ be independent of $y \in \mathbb{R}^n$, $\overline{B}_P(t) = \begin{bmatrix} b_1(t) & \cdots & b_k(t) \end{bmatrix} \in \mathbb{R}^{n \times k}$, where $\{b_1(t), \ldots, b_k(t)\}$ is an orthonormal basis with respect to the standard scalar product on $\mathbb{R}^k$ of $\ker F(y, t)$, and $B_P(t) = \begin{bmatrix} b_{k+1}(t) & \cdots & b_n(t) \end{bmatrix} \in \mathbb{R}^{n \times n-k}$, where $\{b_{k+1}(t), \ldots, b_n(t)\}$ is an orthonormal extension of $\{b_1(t), \ldots, b_k(t)\}$ to an orthonormal basis with respect to the standard scalar product on $\mathbb{R}^n$, $t \in \mathbb{R}$. In addition we assume the domain $\mathcal{D}$ to be so that for each $(y, t) \in \mathcal{D} \times \mathcal{I}$ also $P_B(t) y + s Q_B(t) y \in \mathcal{D}$ for all $s \in [0, 1]$ with $P_B(t) = B_P(t) B_P(t)^\top$ and $Q_B(t) = I - P_B(t)$. Then

$$B_P(t)^\top f(B_P(t) v, t) = 0$$

has a unique solution $v \in \mathbb{R}^k$ and $y = B_P(t) v + u$, with $u \in \ker B_P(t) B_P(t)^\top$ arbitrarily.

Proof. We choose an orthogonal projector $P_B(t)$ along $\ker F(y, t)$ by

$$P_B(t) = B_P(t) B_P(t)^\top.$$

Using Lemma A.7, A.9, the orthogonality of $P_B(t)$ and $\ker F(y, t) = \ker F(y, t)^\top$ yields $\operatorname{im} P_B(t) = \operatorname{im} F(y, t)$ and

$$P_B(t) F(y, t) = F(y, t) = F(y, t) P_B(t)$$

for all $(y, t) \in \mathbb{R}^n \times \mathbb{R}$. The relation

$$f(y, t) = f(P_B(t) y, t)$$

holds true since applying the mean value theorem provides

$$f(y, t) - f(P_B(t) y, t) = \int_0^1 F(sy + (1 - s) P_B(t) y, t)(I - P_B(t)) y \, ds = 0.$$

Due to the choice of $B_P(t)$ and $\overline{B}_P(t)$ the matrix $C(t) = \begin{bmatrix} B_P(t) & \overline{B}_P(t) \end{bmatrix}$ is nonsingular. Hence:

$$f(P_B(t) y, t) = 0 \Leftrightarrow C(t)^\top f(P_B(t) y, t) = 0 \Leftrightarrow \begin{cases} B_P(t)^\top f(P_B(t) y, t) = 0 \\ \overline{B}_P(t)^\top f(P_B(t) y, t) = 0 \end{cases}$$

Notice that $\overline{B}_P(t)^\top f(P_B(t) y, t) = 0$ holds true for all $y \in \mathbb{R}^n$ since

$$\overline{B}_P(t)^\top F(P_B(t) y, t) P_B(t) = 0$$

for all $(y, t) \in \mathbb{R}^n \times \mathcal{I}$, that is, $\overline{B}_P(t)^\top f(P_B(t) y, t)$ is independent of $y \in \mathbb{R}^n$, and due to the requirements that $y \in \mathbb{R}^n$ exists so that $f(y, t) = 0$ for $t \in \mathbb{R}$. Regarding

$$\begin{aligned} 0 &= B_P(t)^\top f(P_B(t) y, t) \\ &= B_P(t)^\top f\left(B_P(t) B_P(t)^\top y, t\right) \\ &= B_P(t)^\top f(B_P(t) v, t) \end{aligned}$$

with $v = B_P(t)^\top y$ and Jacobian given by

$$J(v) = B_P(t)^\top F(B_P(t) v, t) B_P(t).$$

Due to the construction of $B_P(t)$ the Jacobian is nonsingular. $\qquad\square$

Remark 2.60. The orthogonal basis $\{b_1(t), \ldots, b_k(t)\}$, needed in Theorem 2.59 can be calculated by, for example, a QR decomposition.

If we are not interested in a consistent initialization of the DAE (2.30) at $t_0 \in \mathcal{I}$, but finding a solution satisfying the DAE after the first step, we can also apply the implicit Euler method starting with an operating point $y^0 \in \mathcal{M}_0(t_0)$. Since the DAE (2.30) depends linearly on the index-2 components, the approximation obtained at $t_1 = t_0 + h$ is identical to the approximation obtained with a consistent initial value y_0 satisfying $Uy_0 = Uy^0$. In detail, if the implicit Euler method is applied to the DAE (2.30) then the approximation y_1 to $y(t_1)$ is the solution to the following nonlinear system of equations

$$f(y_1) = \frac{1}{h} A(t_1)(d(Uy_1, t_1) - d(Uy_0, t_0)) + \overline{b}(Uy_1, t_1) + B(t_1) Ty_1 \tag{2.49}$$

with $f(y_1) = 0$ and Jacobian $J(Uy_1) = \frac{\partial}{\partial y} f(y_1)$ given by

$$J(Uy_1) = \frac{1}{h} A(t_1) D(Uy_1, t_1) U + \frac{\partial}{\partial Uy} \overline{b}(Uy_1, t_1) U + B(t_1) T.$$

Applying Newton's method to (2.49) yields

$$y_1^1 = y_1^0 - J\left(Uy_1^0\right)^{-1} f\left(y_1^0\right)$$

and

$$J\left(Uy_1^0\right) y_1^1 = J\left(Uy_1^0\right) y_1^0 - f\left(y_1^0\right).$$

That can be reformulated to

$$
\begin{aligned}
J\left(Uy_1^0\right) y_1^1 &= J\left(Uy_1^0\right) y_1^0 - f\left(y_1^0\right) \\
&= \frac{1}{h} A(t_1) D\left(Uy_1^0, t_1\right) Uy_1^0 + \frac{\partial}{\partial Uy} \overline{b}\left(Uy_1^0, t_1\right) Uy_1^0 + B(t_1) Ty_1^0 \\
&\quad - \frac{1}{h} A(t_1) \left(d\left(Uy_1^0, t_1\right) - d(Uy_0, t_0)\right) - \overline{b}\left(Uy_1^0, t_1\right) - B(t_1) Ty_1^0 \\
&= \frac{1}{h} A(t_1) \left(D\left(Uy_1^0, t_1\right) Uy_1^0 - d\left(Uy_1^0, t_1\right) + d(Uy_0, t_0)\right) \\
&\quad + \frac{\partial}{\partial Uy} \overline{b}\left(Uy_1^0, t_1\right) Uy_1^0 - \overline{b}\left(Uy_1^0, t_1\right) \\
&= g\left(Uy_1^0\right)
\end{aligned}
$$

and we obtain

$$y_1^1 = J\left(Uy_1^0\right)^{-1} g\left(Uy_1^0\right).$$

Since this equation depends only on Uy_0, Uy_1^0 and $Uy_0 = Uy^0$, the choice $y_1^0 = y^0$ yields the same approximation at t_1 as the choice $y_1^0 = y_0$. The result does not surprise because it is true for classes of DAEs without a properly stated leading term [Est00]. Consequently, (y_1^1, t_1) is a consistent initial value for the DAE (2.30) at $t_1 \in \mathcal{I}$ provided the rounding errors are zero.

Lemma 2.61. For the DAE (2.30) it is sufficient to start the integration with the implicit Euler using an operating point. All constraints are fulfilled after the first integration step. $\qquad\square$

Remark 2.62. It is a common approach to start the numerical integration with the implicit Euler to overcome the problem of the calculation of a consistent value, but usually it is not proven that the approach works. Here we have shown that starting the integration of the DAE (2.30) using the implicit Euler we only need an operating point and we obtain after one time step a consistent initial value. For this the Theorem 2.58 is essential, since here we prove that starting with an operating point only the index-2 components Ty has to be correct while the non-index-2 components Uy are fixed.

Remark 2.63. The tractability index concept and all techniques presented are not invariant under transformation with respect to solution components which can be chosen freely. Consider the index-1 DAE

$$-x + y = 0 \qquad\qquad \frac{d}{dt}y + 2x = 0$$

where y_0 can be chosen arbitrarily within the tractability index concept, but we can not choose the algebraic component x_0. Inserting the first equation into the second one leads to the index-1 DAE

$$-x + y = 0 \qquad\qquad \frac{d}{dt}x + 2x = 0$$

where x_0 can be chosen freely within the tractability index concept, but we cannot choose the algebraic component y_0. That is dissatisfying since an engineer, for example, does not have the immediate possibility to fix certain initial algebraic components such as the velocity of a car or the energy consumption of an electric device. How to solve this challenging task is still an open issue and may become subject of future research.

For the index-3 or higher DAEs are more of a challenge. If such a DAE is solved by a BDF method, then the solution can have a huge error in the first steps even if consistent initial values are given. Instead of consistent initial values we have to introduce numerically consistent initial values, that is, values fulfilling the numerical constraints to solve the DAE numerically, see [Aré08].

2.3 Summary

In this chapter we have laid the basis of for our later analysis. We have identified problems and challenges in differential-algebraic equation theory and developed and refined methods to tackle them within the tractability index concept for differential-algebraic equations with a properly stated leading term.

The basis was a generalization of the index reduction method by differentiation for index-2 differential-algebraic equations (2.18) to index-1 differential-algebraic equations (2.22) (Lemma 2.41 and 2.43). Next, we have deduced a suitable description of the hidden constraints of the index-2 differential-algebraic equations (2.18) (Theorem 2.45), the local unique solvability (Theorem 2.46) and a perturbation index-2 result (Theorem 2.50). The latter is an important justification for the choice of the tractability index concept in this thesis and is essential to numerics.

To start the numerical integration we focused on consistent initial values for differential-algebraic equations (2.30). For index-2 differential-algebraic equations (2.30) we generalized a step-by-step approach of [Est00] for differential-algebraic equations without a properly stated leading term to differential-algebraic equations with a properly stated leading term. For this we had to calculate an operating point. Next the operating point was corrected by a full rank linear system providing a consistent initialization (Theorem 2.58). For differential-algebraic equations (2.48) we provided a method to compute an operating point (Theorem 2.59). It turned out that for differential-algebraic equations (2.30) it is sufficient to start with an operating point if the implicit Euler method is used for time integration (Lemma 2.61 and Remark 2.62).

3 Maxwell's Equations

Nowadays electric and magnetic fields are an integral part of our technological life. We are surrounded by electric and magnetic fields ranging from induction cooking, mobile phones, wireless networks, electric cars to magnetic resonance tomographs.

The reduction of development costs is a core industrial demand. One way to minimize efforts is to replace as much laboratory testing as possible by numerical simulation predicting the full range of device performance. A common device type is the *electromagnetic*, which is governed by the interaction between electric and magnetic fields fully as described by the partial differential equation system of Maxwell's equations. For the numerical simulation of an electromagnetic device we need to discretize Maxwell's equations in space and time.

A well established method of lines approach for the spatial discretization is the finite integration technique introduced by Thomas Weiland [Wei77] and further developed during the last three decades [MW07]. The finite integration technique is used by our partners in the EU-funded ICESTARS project and the SOFA project, funded by the German government. Moreover, it is successfully applied in established software packages such as MAFIA (Technical University Darmstadt) and CST studio (Computer Simulation Technology AG).

We investigate electromagnetic models described by Maxwell's equations in a potential formulation. They are much used in low and high frequency applications with various formulations and discretizations having already been analyzed, for an overview see [BP89, StM05, Cle05, CMSW11]. Apart from the finite integration technique discretization the cell method [Ton01], particular finite-volume methods [MMS01] and also certain variants of the finite-element method are broadly used [Ned80, Bos98, Göd10]. Here we focus on the finite integration technique discretization scheme. The potential approach results in an adequate problem description that provides a natural link to the concept of potential differences, which are crucial in circuit simulation. However, the potentials are not uniquely defined and to obtain a consistent description a gauge condition is needed, see [Jac98, Bos01]. For the finite integration technique, grad-div formulations based on the Coulomb gauge are well understood, see [CW02, BCDS11]. In this thesis we introduce a new class of gauge conditions in terms of the finite integration technique driven by a Lorenz gauge formulation. After spatial discretization we investigate the structural properties of the resulting differential-algebraic equation formulated with a properly stated leading term. It turns out that the index of the differential-algebraic equation depends on the chosen gauge condition but does not exceed index-2. To concentrate the link to circuit simulation a suitable boundary excitation and current formulation is deduced. Similar differentiation index results are obtained in [BCS12] using a source term excitation and different gauge conditions.

In this chapter the relevant fundamentals of Maxwell's equations are discussed focusing on the basic features of electromagnetism. First, we analyze the electromagnetic fields by using a potential formulation. Different gauge conditions and suitable boundary conditions are discussed. Second, we briefly introduce the finite integration technique. Especially the structural properties of the discrete operators with incorporated boundary conditions are discussed and we introduce a new class of gauge conditions in terms of the finite integration technique. This leads to Maxwell's grid equations and we derive a current formulation and present a boundary excitation for the potential. Third, the resulting differential-algebraic equations are formulated with a properly stated leading term and the new index results are presented, which depend on the chosen gauge condition. In addition, we present an approach to calculate a consistent initialization for Maxwell's grid equations.

3.1 Classical Electromagnetism

Maxwell's equations (ME - Maxwell's Equations) are a set of four coupled partial differential equations postulated by James Clerk Maxwell in the middle of the 19th century and form the basic of the modern theory of electromagnetics (EM - ElectroMagnetic), see [Max64]. These equations describe all phenomena of EM fields by four vector valued functions of space $x \in \Omega \subset \mathbb{R}^3$ and time $t \in \mathcal{I} \subset \mathbb{R}$ on a simple connected domain Ω. The EM quantities are denoted by the *electric* and *magnetic field* $\vec{E}, \vec{H} : \Omega \times \mathcal{I} \to \mathbb{R}^3$ and by the *electric* and *magnetic induction* $\vec{D}, \vec{B} : \Omega \times \mathcal{I} \to \mathbb{R}^3$. An EM field is created by, amongst others, a distribution of electric charges and a current flow. The *distribution of charges* is given by $\rho : \Omega \times \mathcal{I} \to \mathbb{R}^3$ while the *conduction current density* is described by $\vec{J}_c : \Omega \times \mathcal{I} \to \mathbb{R}^3$, see [Jac98, HW05].

Today ME in differential form reads:

$$\nabla \cdot \vec{D} = \rho \tag{3.1}$$

$$\nabla \cdot \vec{B} = 0 \tag{3.2}$$

$$\nabla \times \vec{E} = -\frac{\partial}{\partial t}\vec{B} \tag{3.3}$$

$$\nabla \times \vec{H} = \vec{J}_c + \frac{\partial}{\partial t}\vec{D} \tag{3.4}$$

These equations describe the spatial and temporal behavior of the EM quantities. Tables of SI units are given in Table 3.1 and 3.2.

quantity	SI units
$\vec{E}$	V/m
$\vec{D}$	$C/m^2 = As/m^2$
$\vec{B}$	$T = Vs/m^2$
$\vec{H}$	A/m
$\vec{J}_c, \vec{J}_d, \vec{J}_t$	A/m^2
φ	V
$\vec{A}$	Wb/m = Vs/m
$\vec{\Pi}$	V/m
ρ	$C/m^3 = As/m^3$

operator	SI units
$\frac{\partial}{\partial t}$	1/s
∇	1/m
$\nabla\times$	1/m
$\nabla\cdot$	1/m

Table 3.1: Field quantities.   Table 3.2: Differential operators.

ME are the work of several well-known physicians. That is why the individual equations are attributed to other scientists. But Maxwell grouped all the equations together into a consistent set and introduced the displacement current. The *Gauss' law* (3.1) describes the effect of the charge density on the electric induction and *Gauss' law for magnetism* (3.2) expresses the fact that magnetic induction is solenoidal. The *Maxwell-Faraday's law* (3.3) describes the effect of a time changing magnetic field on the electric field. Finally, *Maxwell-Ampère's law* (3.4) gives the effect of the total current density on the magnetic field. The *total* and *displacement current density* $\vec{J}_t, \vec{J}_d : \Omega \times \mathcal{I} \to \mathbb{R}^3$ are given by

$$\vec{J}_t = \vec{J}_c + \vec{J}_d \text{ and } \vec{J}_d = \frac{\partial}{\partial t}\vec{D}.$$

An essential feature of ME is that electric charges are conserved. For this we derive the charge-current *continuity equation* from ME. The divergence of (3.4) and the time derivative of (3.1) lead to the continuity equation

$$\nabla \cdot \vec{J}_c + \frac{\partial}{\partial t}\rho = 0 \tag{3.5}$$

expressing the conservation of electric charges. The continuity equation reveals that ME are not independent. If charge is conserved, then Gauss' law and Gauss' law for magnetism are consequences of Maxwell-Faraday's law and Maxwell-Ampère's law. Taking the divergence of (3.3) and (3.4) and interchanging the derivatives we obtain

$$\frac{\partial}{\partial t}\nabla \cdot \vec{B} = 0 \text{ and } \frac{\partial}{\partial t}\left(\nabla \cdot \vec{D} - \rho\right) = 0$$

using the continuity equation (3.5). Thus, if the divergence conditions (3.1) and (3.2) are fulfilled at one time they hold for all time, see [Mon03]. Hence the divergence conditions are consequences of the dynamic curl conditions (3.3) and (3.4) and can be seen as restriction on valid initial conditions for the Maxwell-Faraday's law and Maxwell-Ampère's law. Therefore we conclude that the whole time evolution is completely specified by the dynamic curl conditions. ME are completed by three *constitutive laws*. The laws relate

quantity	SI units
ε	F/m=As/Vm
ν	m/H=Am/Vs
σ	S/m=A/Vm

Table 3.3: Material properties.

$\vec{E}$ and $\vec{B}$ to $\vec{D}$, $\vec{J}_c$ and $\vec{H}$. These laws depend on the material properties in the domain occupied by the EM field. One distinguishes between linear and nonlinear, homogeneous and inhomogeneous, isotropic and anisotropic materials. For linear materials the constitutive laws are independent of the fields quantities. The constitutive laws of nonlinear materials depend on the fields quantities. For homogeneous materials the constitutive laws are independent on the spatial coordinates. The constitutive laws of inhomogeneous materials are functions of the spatial coordinates. Isotropic and anisotropic materials are characterized by the absence or presence of a dependence of the constitutive laws upon the spatial direction, see [Ben06].

We restrict ourselves to the following constitutive laws. The first constitutive law relates $\vec{E}$ and $\vec{D}$ by

$$\vec{D} = \varepsilon\vec{E}, \tag{3.6}$$

with $\varepsilon : \Omega \to \mathbb{R}$ and the *permittivity* ε depending on the spatial coordinates only. The second constitutive law relates $\vec{B}$ and $\vec{H}$ by

$$\vec{H} = \nu(\vec{B})\vec{B}, \tag{3.7}$$

with $\nu : \Omega \times \mathbb{R}^3 \to \mathbb{R}^{3\times3}$ and the *reluctivity* ν depending on the spatial coordinates and depending nonlinearly and anisotropically on the magnetic induction. The reluctivity is the inverse of the *permeability* μ.

In case of conductive materials the electric field $\vec{E}$ itself gives rise to a current flow. That leads to the last constitutive law also known as *Ohm's law*. As long as the field strengths are not too large we can assume that Ohm's law is fulfilled. It relates $\vec{E}$ and $\vec{J}_c$ by

$$\vec{J}_c = \sigma\vec{E}, \tag{3.8}$$

with $\sigma : \Omega \to \mathbb{R}$ and the *conductivity* σ depending on the spatial coordinates only. In insulating materials we can assume that σ vanishes.

Assumption 3.1 (constitutive laws). The materials have:

(i) Linear, inhomogeneous and isotropic permittivity and conductivity.

(ii) Nonlinear, inhomogeneous and anisotropic reluctivity given by *Brauer's model*, see [Sch11, BH91].

Table 3.3 shows the corresponding SI units of the material properties. In this thesis we restrict ourselves to materials fulfilling Assumption 3.1.

Classification of Electromagnetic Problems

The behavior of EM fields is governed by ME. To simplify the calculation of ME there are several approaches that disregard effects depending on the speed of propagation of the EM waves. Common simplifications are:

(i) Static fields: The time dependence in the EM quantities are neglected, that is, $\frac{\partial}{\partial t}\vec{B} = 0$ and $\frac{\partial}{\partial t}\vec{D} = 0$.

(ii) *Magnetoquasistatic* (MQS - <u>M</u>agneto<u>Q</u>uasi<u>S</u>tatic): The electric induction $\vec{D}$ is slowly varying and the time dependence is neglected, that is, $\frac{\partial}{\partial t}\vec{D} = 0$.

(iii) *Electroquasistatic*: The magnetic induction $\vec{B}$ is slowly varying and the time dependence is neglected, that is, $\frac{\partial}{\partial t}\vec{B} = 0$.

Every simplification has an impact on the solution, that is, we have to take care if a simplification is really admissible, see [HM89]. In this thesis we mainly focus on ME without simplification, that is, we consider the "full set" of ME in time domain.

3.1.1 Potential Formulation and Gauge Conditions

When studying ME it is often convenient to introduce auxiliary functions that simplify the representation of ME. For our investigations we use a potential approach, see [BP89, Jac98, StM05, HW05].

From Gauss' law for magnetism (3.2) we deduce from Helmholtz decomposition that there is a vector field $\vec{A} : \Omega \times \mathcal{I} \to \mathbb{R}^3$ such that

$$\vec{B} = \nabla \times \vec{A}$$

and using Maxwell-Faraday's law (3.3) we obtain

$$\nabla \times \left(\vec{E} + \frac{\partial}{\partial t}\vec{A} \right) = 0.$$

Thus using Helmholtz decomposition we can conclude that there is a scalar function
$\varphi : \Omega \times \mathcal{I} \to \mathbb{R}$ such that

$$\vec{E} = -\nabla \varphi - \frac{\partial}{\partial t}\vec{A}.$$

This potential approach is the so-called $(\vec{A}, \varphi)$-formulation with the *vector potential* $\vec{A}$ and *scalar potential* φ, see [StM05]. Note there are different potential approaches and for an overview we refer to Tabelle 2.4 in [Koc09]. The $(\vec{A}, \varphi)$-formulation has the advantage that the scalar potential φ provides a natural link to the concept of potential differences which plays a crucial role in conventional simulations of electric circuits. A second advantage is that Gauss' law for magnetism and Maxwell-Faraday's law are automatically fulfilled. A visual representation of all quantities is given in Figure 3.1(a).

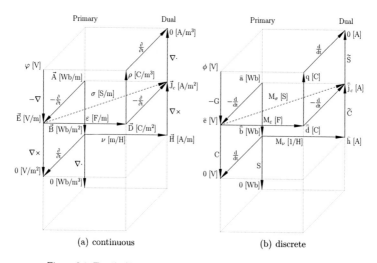

(a) continuous
(b) discrete

Figure 3.1: Tonti's diagram or *Maxwell's house*, [Ton95, Cle05, StM05].

The potential approach has a drawback: The scalar and vector potentials exhibit a so-called *gauge freedom*, that is, there are arbitrary in the sense that $\vec{B}$ and $\vec{E}$ are left unchanged if the *gauge transformation*

$$\vec{A}' = \vec{A} + \nabla \chi \text{ and } \varphi' = \varphi - \frac{\partial}{\partial t}\chi$$

is applied, where the *gauge function* $\chi : \Omega \times \mathcal{I} \to \mathbb{R}$ is an arbitrary scalar function, see [Jac98, HW05]. For $\vec{B}$ and $\vec{E}$ we have

$$\vec{B}' = \nabla \times \vec{A}'$$

$$= \nabla \times \vec{A} + \nabla \times \nabla \chi$$
$$= \nabla \times \vec{A}$$
$$= \vec{B}$$

and

$$\vec{E}' = -\nabla \varphi' - \frac{\partial}{\partial t} \vec{A}'$$
$$= -\nabla \varphi - \nabla \frac{\partial}{\partial t} \chi - \frac{\partial}{\partial t} \vec{A} + \frac{\partial}{\partial t} \nabla \chi$$
$$= -\nabla \varphi - \frac{\partial}{\partial t} \vec{A}$$
$$= \vec{E}.$$

A physical law which does not change under a gauge transformation is said to be *gauge invariant*. In that sense $\vec{E}$ and $\vec{B}$ are gauge invariant. To obtain a unique solution the next step is to remove the gauge freedom of $\vec{A}$ and φ. For that reason we fix a gauge function except for a constant scalar field by choosing a *gauge condition*. In the following we introduce the two most common gauge conditions, namely the *Coulomb gauge* and *Lorenz gauge* given by

$$\nabla \cdot \vec{A} = 0 \tag{3.9}$$

and

$$\varepsilon \mu \frac{\partial}{\partial t} \varphi + \nabla \cdot \vec{A} = 0 \tag{3.10}$$

for the case of linear, homogeneous and isotropic materials.

Remark. Lorenz gauge is named after Ludvig Lorenz. It is an invariant condition, and is often wrongly called Lorentz gauge because of confusing with Hendrik Lorentz, after whom Lorentz covariance is named.

To show the impact of the two gauge conditions we assume ε and μ to be constant and the functions ρ and $\vec{J}_c$ to be given, but related by the continuity equation (3.5), see [MRT05]. That is, we regard $\vec{J}_c$ as a given source current density. Then Gauss' law (3.1), Maxwell-Ampère's law (3.4) and the constitutive laws (3.6) and (3.7) lead to

$$\Delta \varphi + \frac{\partial}{\partial t} \nabla \cdot \vec{A} = -\frac{1}{\varepsilon} \rho$$
$$\nabla^2 \vec{A} - \varepsilon \mu \frac{\partial^2}{\partial t^2} \vec{A} - \nabla \left(\nabla \cdot \vec{A} + \varepsilon \mu \frac{\partial}{\partial t} \varphi \right) = -\mu \vec{J}_c \tag{3.11}$$

where the vector Laplace operator is denoted by ∇^2. Applying Coulomb gauge to (3.11) we get the semi-decoupled system

$$\Delta\varphi = -\frac{1}{\varepsilon}\rho$$
$$\nabla^2\vec{A} - \varepsilon\mu\frac{\partial^2}{\partial t^2}\vec{A} = -\mu\vec{J}_c + \varepsilon\mu\frac{\partial}{\partial t}\nabla\varphi \tag{3.12}$$

consisting of an elliptic equation for φ and three wave equations for $\vec{A}$. Applying Lorenz gauge to (3.11) we can deduce the fully decoupled system

$$\Delta\varphi - \varepsilon\mu\frac{\partial^2}{\partial t^2}\varphi = -\frac{1}{\varepsilon}\rho$$
$$\nabla^2\vec{A} - \varepsilon\mu\frac{\partial^2}{\partial t^2}\vec{A} = -\mu\vec{J}_c \tag{3.13}$$

consisting of four wave equations.

In both cases the systems lead to an unique solution if we choose initial and boundary conditions properly, see [Eva10], and hence the gauge function χ is fixed. To derive the system (3.12) and (3.13) we apply Coulomb and Lorenz gauge directly to (3.11). To make sure that the gauge conditions are fulfilled implicitly we need to choose the initial and boundary conditions such that we obtain only the trivial solution for the homogeneous wave equation

$$\Delta\psi - \varepsilon\mu\frac{\partial^2}{\partial t^2}\psi = 0 \tag{3.14}$$

with $\psi : \Omega \times \mathcal{I} \to \mathbb{R}$ given by $\psi = \nabla \cdot \vec{A}$ and $\psi = \varepsilon\mu\frac{\partial}{\partial t}\varphi + \nabla \cdot \vec{A}$, depending on the chosen gauge condition. We obtain (3.14) by taking the continuity equation (3.5) into account. That is important since the systems (3.12) and (3.13) solve ME if and only if the applied gauge is implicitly fulfilled. Note that both gauges regularize the curl-curl operator in the sense that a Green function exists to determine the vector potential $\vec{A}$ uniquely.

For our later analysis we need to generalize the Coulomb and Lorenz gauge to obtain suitable gauge conditons for the spatial discretization method presented in the next section. We rewrite Maxwell-Ampère's law (3.4) to

$$\vec{J}_c = \left(\nabla \times \nu\nabla \times \vec{A} - \zeta\nabla\xi\nabla \cdot \zeta\vec{A}\right) + \left(\varepsilon\nabla\frac{\partial}{\partial t}\varphi + \zeta\nabla\xi\nabla \cdot \zeta\vec{A}\right) + \frac{\partial^2}{\partial t^2}\varepsilon\vec{A}$$

with *artificial material properties* $\zeta, \xi : \Omega \times \mathbb{R} \to \mathbb{R}$ such that the SI units of ν and $\zeta^2\xi$ match. A possible *class of gauge conditions* reads

$$\vartheta\varepsilon\nabla\frac{\partial}{\partial t}\varphi + \zeta\nabla\xi\nabla \cdot \zeta\vec{A} = 0 \tag{3.15}$$

with $\vartheta \in \mathbb{R}$. For $\vartheta = 0$ we obtain a grad-div type of Coulomb gauge

$$\zeta \nabla \xi \nabla \cdot \zeta \vec{A} = 0.$$

Moreover for $\vartheta = 1$ we get a type of Lorenz gauge

$$\varepsilon \nabla \frac{\partial}{\partial t} \varphi + \zeta \nabla \xi \nabla \cdot \zeta \vec{A} = 0.$$

For further discussion on gauge conditions we refer to [BCS12, StM05, Bos01, CMSW11].

Finally, ME formulated in terms of potentials using Maxwell-Ampère's law (3.4), the constitutive laws (3.6), (3.7) and (3.8), and the gauge condition (3.15) reads

$$\vartheta \varepsilon \nabla \frac{\partial}{\partial t} \varphi + \zeta \nabla \xi \nabla \cdot \zeta \vec{A} = 0$$

$$\nabla \times \nu \nabla \times \vec{A} + \varepsilon \frac{\partial}{\partial t} \left(\nabla \varphi + \vec{\Pi} \right) + \sigma \left(\nabla \varphi + \vec{\Pi} \right) = 0 \qquad (3.16)$$

$$\frac{\partial}{\partial t} \vec{A} - \vec{\Pi} = 0$$

utilizing an *auxiliary vector field* $\vec{\Pi} : \Omega \times \mathcal{I} \to \mathbb{R}^3$ to avoid the second-order differentiation in time for $\vec{A}$.

3.1.2 Boundary and Interface Conditions

In general, EM field problems are not restricted, open boundary problems. However, for our later investigations we have to restrict ourselves to a finite domain $\Omega \subsetneq \mathbb{R}^3$. That is, we deal with an artificially bounded problem.

Assumption 3.2. The finite domain $\Omega \subsetneq \mathbb{R}^3$ is simply connected with the boundary $\Gamma = \partial \overline{\Omega}$.

In case of MQS the truncation of the domain is reasonable if a sufficiently large region of air is around the MQS device, since the magnetic induction decays rapidly in the air towards the boundary. As a general rule it recommends the distance from the device to the boundary to be at least five times the radius of the device, see [CK97].

Remark 3.3. A MQS device is an EM device under the MQS assumption.

Unfortunately, this argumentation is not valid in our case since we will assume that the device is connected to the boundary. We assume that the main part of the device is sufficiently far away from the boundary and that wires with a good conductivity connect the main part of the device to the boundary. Due to the damped wave equations character of ME the fields decays towards the boundary.

To complete the system (3.16) we need boundary conditions. In addition, we have to handle discontinuities of ε, ν and σ which can appear at the boundary between different materials in our bounded domain Ω. We denote by Γ_{int} the *internal boundary* between different materials. The conditions on the internal boundary are called interface conditions.

We consider an internal boundary separating two different materials 1 and 2 with material properties $(\varepsilon_1, \sigma_1, \nu_1)$ and $(\varepsilon_2, \sigma_2, \nu_2)$. Interface conditions are obtained by applying the Gauss' theorem and Stokes' theorem to ME in a small region at the internal boundary between two materials:

$$\left(\vec{E}_2 - \vec{E}_1 \right) \cdot \vec{n}_{\parallel} = 0$$

$$\left(\vec{B}_2 - \vec{B}_1 \right) \cdot \vec{n}_{\perp} = 0$$

$$\left(\vec{D}_2 - \vec{D}_1 \right) \cdot \vec{n}_{\perp} = \varrho$$

$$\left(\vec{H}_2 - \vec{H}_1 \right) \cdot \vec{n}_{\parallel} = \kappa$$

A detailed derivation is given in [Jac98, Str07]. The subscripts 1 and 2 of the EM quantities denote the quantities in materials 1 and 2. Here ϱ describes the surface charge density and κ the surface current density with

$$\varrho, \kappa : \Gamma_{int} \to \mathbb{R}.$$

That is, the tangential component of $\vec{E}$ and the normal component of $\vec{B}$ are continuous functions across the internal boundary. The method to derive the interface conditions is known as *pill-box method*.

The interface conditions motivates boundary conditions for $\vec{E}$ and $\vec{B}$. One approach is the so-called *electric boundary condition* (PEC - Perfectly Electric Conducting) and are also called "flux wall" or "current gate" boundary conditions, see [Cle98, Ben06]. We assume:

$$\vec{E} \cdot \vec{n}_{\parallel} = 0 \tag{3.17}$$

$$\vec{B} \cdot \vec{n}_{\perp} = 0 \tag{3.18}$$

The idea is to think of a complete universe, where ME are also true outside the simulation domain Ω. The picture is to attached a perfect conductor from outside at the boundary, where the magnetic induction does not pass through.

The next step is to interpret and motivate the boundary conditions for $\vec{E}$ and $\vec{B}$ in terms of the potentials $\vec{A}$ and φ.

Assumption 3.4. We assume that the boundary consists of $k \in \mathbb{N}$ disjoint parts with $\Gamma = \bigcup_{i=1}^{k} \Gamma_i$ and for every Γ_i the material properties $(\varepsilon, \sigma, \nu)$ to be constant.

Let Assumption 3.4 be fulfilled. We examine an arbitrarily boundary part Γ_i. Due to (3.18) we get

$$0 = \vec{B} \cdot \vec{n}_\perp = \nabla \times \vec{A} \cdot \vec{n}_\perp$$

and we deduce that

$$\vec{A} \cdot \vec{n}_\shortparallel = 0 \qquad (3.19)$$

is a possible choice. Next we inspect (3.17) and we obtain

$$0 = \vec{E} \cdot \vec{n}_\shortparallel = -\nabla \varphi \cdot \vec{n}_\shortparallel - \frac{\partial}{\partial t}\vec{A} \cdot \vec{n}_\shortparallel = -\nabla \varphi \cdot \vec{n}_\shortparallel.$$

To fulfill that condition a possible choice is

$$\nabla \varphi \cdot \vec{n}_\shortparallel = 0. \qquad (3.20)$$

The conditions (3.19) and (3.20) can be interpreted as Dirichlet boundary conditions for the potentials.

Applying the pill-box method using the gauge conditions (3.9) and (3.10) it is possible to show that $\nabla \varphi$ inherits the discontinuity of $\vec{E}$ at internal boundaries and $\vec{A}$ is continuous, see [AH01]. This motivates choosing homogeneous Dirichlet boundary conditions for $\vec{A}$ and spatial-constant time-dependent Dirichlet boundary conditions for φ on each Γ_i. In addition, we choose Dirichlet boundary conditions for $\vec{\Pi}$ in accordance with $\vec{A}$. This set of boundary conditions are a suitable link to circuit simulation, where the boundary condition for the scalar potential φ can be identified with the applied potentials at the device contacts.

Essential for our later analysis of ME is the charge conservation expressed by the continuity equation (3.5), since including EM devices into circuit models are only possible if charges are conserved. Due to the definition of the total current $\vec{J}_t = \vec{J}_c + \vec{J}_d$ we obtain

$$\int_\Gamma \vec{J}_t \cdot \vec{n}_\perp dF = \int_\Omega \nabla \cdot \vec{J}_t dV = 0 \text{ with } \vec{J}_t = \nabla \times \nu \nabla \times \vec{A},$$

that is, the sum of in- and outgoing currents equals. Without loss of generality we suppose that we number the disjoint boundary parts Γ_i such that the first $n_E < k$ boundary parts have the material property $\sigma \neq 0$ and the last $k - n_E$ boundary parts have the material property $\sigma = 0$. We call $\Gamma_g = \bigcup_{i=n_E+1}^{k} \Gamma_i$ the *mass contact* while the other Γ_i are called *conductive contacts* and we get

$$j_g = -\sum_{i=1}^{n_E} j_i \text{ with } j_i = \int_{\Gamma_i} \vec{J}_t \cdot \vec{n}_\perp dF.$$

This means that the current j_g flowing through the mass contact of the EM device is the negative sum of the currents j_i, $i = 1, \ldots, n_E$, flowing through the conductive contacts. The picture is that each conductive contact is connected to a wire from outside while the mass contact is grounded.

3.2 Finite Integration Technique

This section provides a survey on the *finite integration technique* (FIT - Finite Integration Technique) for spatial discretization for solving ME in integral form. That approach was developed and formulated by Thomas Weiland [Wei77] and is based on a staggered discretization. For orthogonal grids in time domain the FIT is equivalent to the finite-difference time-domain-scheme of Kane Yee, also known as leap-frog scheme, see [Yee66].

The first step in the FIT discretization is the decomposition of the domain Ω into a finite number of three-dimensional volumes so that the intersection of two different volumes is either empty - or a two-dimensional facet, a one-dimensional edge or a zero-dimensional node shared by both volumes. This decomposition yields a finite volumes complex $\mathcal{G}$. To each edge of the volumes we prescribe an initial orientation, so that $\mathcal{G}$ can be characterized as a directed graph, see Chapter B for the notation in Graph theory. The volume facets are supplied with an initial orientation, too.

For a rectilinear grid in Cartesian coordinates on a brick-shaped domain Ω, see [Wei77, TW96], the corresponding volumes complex $\mathcal{G}$ reads

$$\mathcal{G} = \left\{ V(n) = V\left(n\left(i_x, i_y, i_z\right)\right) \,\middle|\, V\left(n\left(i_x, i_y, i_z\right)\right) = \left[x_{i_x}, x_{i_x+1}\right] \times \left[y_{i_y}, y_{i_y+1}\right] \times \left[z_{i_z}, z_{i_z+1}\right], \right.$$
$$\left. i_x = 1, \ldots, N_x - 1, i_y = 1, \ldots, N_y - 1, i_z = 1, \ldots, N_z - 1 \right\}$$

where N_x, N_y and N_z are the total numbers of (grid) nodes in x-, y- and z- direction, respectively. The total number of nodes is then $N = N_x N_y N_z$. The space indices i_x, i_y and i_z can be reduced to one canonical space index

$$n = n\left(i_x, i_y, i_z\right) = 1 + \left(i_x - 1\right) K_x + \left(i_y - 1\right) K_y + \left(i_z - 1\right) K_z \leqslant N$$

where $K_x = 1$, $K_y = N_x$, $K_z = N_x N_y$ and $i_x = 1, \ldots, N_x$, $i_y = 1, \ldots, N_y$, $i_z = 1, \ldots, N_z$.

To each node $N(n)$ we associate three (grid) edges $E_x(n)$, $E_y(n)$, $E_z(n)$, three (grid) facets $F_x(n)$, $F_y(n)$, $F_z(n)$ and finally, one (grid) volume $V(n)$.
The orientation of edges and facets is given as follows: The front node of the edge $E_w(n)$ in w-direction is $N(n)$. A facet $F_w(n)$ is defined by the lower left node $N(n)$ and the direction w, in which its normal vector points.

Remark 3.5. The numbering scheme of the grid $\mathcal{G}$ introduces phantom edges, facets and volumes at the boundary of the finite domain Ω. To not disrupt the convenient numbering scheme, we tackle this issue later, see Subsection 3.2.4.

The FIT makes use of two staggered grids. The primary grid $\mathcal{G}$ is supported by a dual grid $\tilde{\mathcal{G}}$, which is constructed by connecting the center points of neighboring primary volumes sharing a facet, see Figure 3.2. The center points define the dual (grid) nodes $\tilde{N}(n)$. The definition of the dual (grid) edges $\tilde{E}_w(n)$, facets $\tilde{F}_w(n)$ and volume $\tilde{V}(n)$, are analogous to the primary grid. The orientation of dual edges and dual facets is given

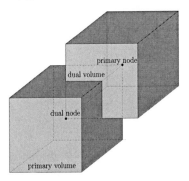

Figure 3.2: Spatial allocation of a primary cell and a dual cell of the grid doublet, see [CW01b].

as follows: The dual back node of the dual edge $\tilde{E}_w(n)$ in w-direction is $\tilde{N}(n)$. A dual facet $\tilde{F}_w(n)$ is defined by the upper right dual node $\tilde{N}(n)$ and the direction w, in which its normal vector points.

Remark 3.6. With this definition of the dual grid it is ensured that there is a one-to-one relation between nodes and edges of $\mathcal{G}$ and volumes and facets of $\tilde{\mathcal{G}}$ and vice versa.

The collection of all primary nodes and primary edges are denoted by $\mathcal{N}$ and $\mathcal{E}$.

3.2.1 Maxwell's Grid Equations

The formulation of discrete approach to electromagnetism arises from the mapping of ME in their integral form and the constitutive laws on $\{\mathcal{G}, \tilde{\mathcal{G}}\}$. As variables of the FIT we introduce electric and magnetic voltages located on the edges defined by

$$\widehat{e}_w(n) = \int_{E_w(n)} \vec{E} \cdot \vec{n}_{\shortparallel} dE, \; \widehat{h}_w(n) = \int_{\tilde{E}_w(n)} \vec{H} \cdot \vec{n}_{\shortparallel} dE,$$

as well as magnetic and electric fluxes and electric currents allocated at the facets defined by

$$\widehat{b}_w(n) = \int_{F_w(n)} \vec{B} \cdot \vec{n}_{\perp} dF, \; \widehat{d}_w(n) = \int_{\tilde{F}_w(n)} \vec{D} \cdot \vec{n}_{\perp} dF, \; \widehat{j}_{c,w}(n) = \int_{\tilde{F}_w(n)} \vec{J}_c \cdot \vec{n}_{\perp} dF.$$

The variables are tagged by arcs according to the underlying geometric object, see [Bos88]. For a convenient notation we introduce the state variable vector

$$\widehat{e} = (\widehat{e}_x(1), \ldots, \widehat{e}_x(N), \widehat{e}_y(N), \ldots, \widehat{e}_y(N), \widehat{e}_y(1), \ldots, \widehat{e}_y(N))$$

and the vector $\widehat{h}$, $\widehat{\widehat{b}}$, $\widehat{\widehat{d}}$ and $\widehat{\widehat{j}}_c$ are defined analogously. This notation allows to write ME in terms of the FIT discretization. Gauss' law for magnetism (3.2), for example,

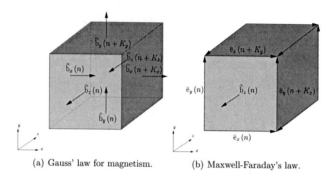

(a) Gauss' law for magnetism.   (b) Maxwell-Faraday's law.

Figure 3.3: Allocation of the FIT degrees of freedom on the primary grid.

integrated over a volume $V(n)$, see Figure 3.3(a), can be written as

$$-\widehat{\widehat{b}}_x(n) + \widehat{\widehat{b}}_x(n + K_x) - \widehat{\widehat{b}}_y(n) + \widehat{\widehat{b}}_y(n + K_y) - \widehat{\widehat{b}}_z(n) + \widehat{\widehat{b}}_z(n + K_z) = 0$$

using Gauss' theorem. The relations for all volumes are collected in the equation

$$\underbrace{\begin{bmatrix} & \vdots & \vdots & & \vdots & & \vdots & & \vdots & & \vdots & \\ \cdots & -1 & 1 & \cdots & -1 & \cdots & 1 & \cdots & -1 & \cdots & 1 \cdots \\ & \vdots & \vdots & & \vdots & & \vdots & & \vdots & & \vdots & \end{bmatrix}}_{=S} \begin{pmatrix} \vdots \\ \widehat{\widehat{b}}_x(n) \\ \widehat{\widehat{b}}_x(n + K_x) \\ \vdots \\ \widehat{\widehat{b}}_y(n) \\ \vdots \\ \widehat{\widehat{b}}_y(n + K_y) \\ \vdots \\ \widehat{\widehat{b}}_z(n) \\ \vdots \\ \widehat{\widehat{b}}_z(n + K_z) \\ \vdots \end{pmatrix} = 0.$$

The Maxwell-Faraday's law (3.3) integrated over a volume $F_z(n)$, see Figure 3.3(b), leads to

$$\widehat{e}_x(n) - \widehat{e}_x(n + K_y) - \widehat{e}_y(n) + \widehat{e}_y(n + K_x) = \frac{d}{dt}\widehat{\widehat{b}}_z(n)$$

using Stokes' theorem and can be organized for all facets by

$$\underbrace{\begin{bmatrix} & \vdots & & \vdots & & \vdots & & \vdots & \\ \cdots & -1 & \cdots & 1 & \cdots & 1 & & -1 & \cdots \\ & \vdots & & \vdots & & \vdots & & \vdots & \end{bmatrix}}_{=C} \begin{pmatrix} \vdots \\ \widehat{e}_x(n) \\ \vdots \\ \widehat{e}_x(n + K_y) \\ \vdots \\ \widehat{e}_y(n) \\ \widehat{e}_y(n + K_x) \\ \vdots \end{pmatrix} = \frac{d}{dt}\begin{pmatrix} \vdots \\ \widehat{\widehat{b}}_z(n) \\ \vdots \end{pmatrix}.$$

The discretization of both laws exploits the numbering scheme and we refer to [Wei77, Wei96] for more details. The discretization of Gauss' law (3.1) and Maxwell-Ampère's law (3.4) is analogously to the procedure described above with the only difference that the discrete quantities are allocated at the dual grid elements. Finally, the FIT has

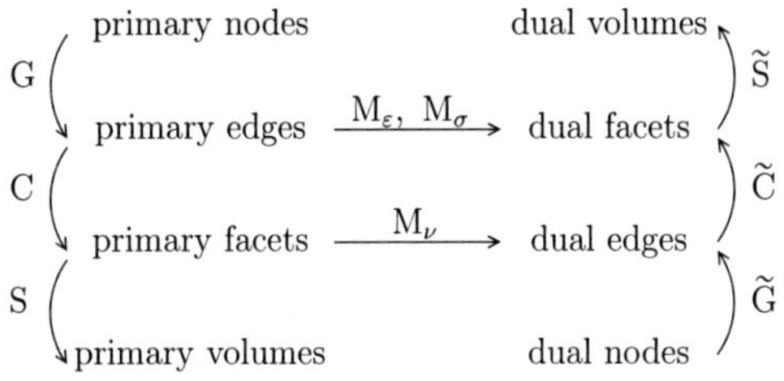

Figure 3.4: Operator mapping.

translated ME exactly into *Maxwell's grid equations* (MGE - Maxwell's Grid Equations), [CW01b], given by

$$\widetilde{S}\widehat{\widehat{d}} = q \tag{3.21}$$

$$S\widehat{b} = 0 \tag{3.22}$$

$$C\widehat{e} = -\frac{d}{dt}\widehat{b} \tag{3.23}$$

$$\widetilde{C}\widehat{h} = \frac{d}{dt}\widehat{\widehat{d}} + \widehat{\widehat{j}}_c \tag{3.24}$$

with the *discrete Gauss' law* (3.21), *discrete Gauss' law for magnetism* (3.22), the *discrete Maxwell-Faraday's law* (3.23) and the *discrete Maxwell-Ampère's law* (3.24). The *discrete curl operators* C and $\tilde{C}$, the *discrete divergence operators* S and $\tilde{S}$ on the primal and dual grid, respectively. The discrete curl operators contain only information on the incidence relation of the volume edges and on their orientation, see Figure 3.5. The divergence operators collect information on the incidence relation and on the orientation of the facets of the volumes. The curl operator maps from edges to facets and the divergence operator from facets to volumes, see Figure 3.4. The unknowns are the *discrete electric field* and *magnetic field* $\widehat{e}, \widehat{h} : \mathcal{I} \to \mathbb{R}^{3N}$, *discrete electric* and *magnetic induction* $\widehat{\widehat{d}}, \widehat{\widehat{b}} : \mathcal{I} \to \mathbb{R}^{3N}$, *conduction current* $\widehat{\widehat{j}}_c : \mathcal{I} \to \mathbb{R}^{3N}$ and *distribution of charges* q $: \mathcal{I} \to \mathbb{R}^N$. Tables of SI units are given in Table 3.4 and 3.5.

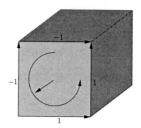

Figure 3.5: Orientation of the curl.

Remark 3.7. The discrete distribution of charges q is located on dual volumes and hence q should be written as $\widehat{\widehat{q}}$ to be consistent with the notation. Nonetheless, for clarity, we simple write q in abuse of notation.

So far the discretization of the physical laws does not require any approximation since the ME have been directly applied to the grid by using topological information only. For a complete discretization of ME the constitutive laws (3.6), (3.7) and (3.8) have to be related to the discrete EM quantities allocated at the grid doublet. At this point all

quantity	geometric object	SI units
$\widehat{e}$	primary edges	V
$\widehat{\widehat{d}}$	dual surfaces	C=As
$\widehat{b}$	primary surfaces	Wb=Vs
$\widehat{h}$	dual edges	A
$\widehat{\widehat{j}}_c, \widehat{\widehat{j}}_t$	dual surfaces	A
ϕ	primary node	V
$\widehat{a}$	primary edges	Wb=Vs
$\widehat{\pi}$	primary edges	V
q	dual volumes	C=As

Table 3.4: Discrete field quantities.

operator	SI units
$\frac{d}{dt}$	1/s
G, $\tilde{G}$	1
C, $\tilde{C}$	1
S, $\tilde{S}$	1

Table 3.5: Discrete differential operators.

metric information enters the spatial discretization and the constitutive laws establish a coupling between the primary and the dual EM quantities. Now the need for the grid doublet becomes clear. For example the discrete version of the constitutive laws (3.6) needs to relate $\widehat{e}$ and $\widehat{d}$, but these discrete quantities are defined on different geometric objects. We can relate them because of the one-to-one relation between edges of $\mathcal{G}$ and facets of $\widetilde{\mathcal{G}}$.

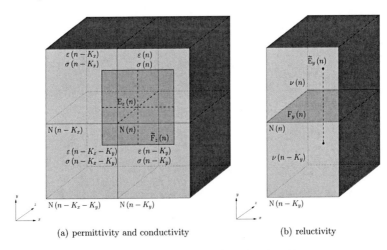

(a) permittivity and conductivity   (b) reluctivity

Figure 3.6: Material properties located on the grid.

Assumption 3.8. The material properties are constant in each primary volume.

Let Assumption 3.1 and 3.8 be true. To derive a discrete version of the constitutive laws (3.6) for linear, inhomogeneous and isotropic permittivities we employ the rectangle rule. Regarding the edge $\mathrm{E}_z(n)$ and the facet $\widetilde{\mathrm{F}}_z(n)$. Using the midpoint rectangle rule we get

$$\widehat{e}_z(n) = |\mathrm{E}_z(n)| \, |\vec{\mathrm{E}}|_{z,n} + \mathcal{O}\left(h^3\right) \tag{3.25}$$

where $|\vec{\mathrm{E}}|_{z,n}$ is the sample value of the electric field at the midpoint of the edge $\mathrm{E}_z(n)$, $|\mathrm{E}_z(n)|$ is the edge lengths and

$$h = \max_{\substack{w \in \{x,y,z\} \\ 1 \leqslant n \leqslant N}} |\mathrm{E}_w(n)|$$

is the maximum length of the edges. The discontinuities of electric induction $\vec{\mathrm{D}}$ at internal boundaries in normal direction does not effect the discretization since we need to switch to the electric field $\vec{\mathrm{E}}$ for the discretization of the constitutive law (3.6). Applying

quantity	SI units
M_ε	F=As/V
M_ν	1/H=A/Vs
M_σ	S=A/V

Table 3.6: Discrete material matrices.

the top-left and top-right rectangle rule we get

$$\widehat{\widehat{d}}_z(n) = \widetilde{\varepsilon}_z(n) |\vec{E}|_{z,n} + \mathcal{O}\left(h^3\right) \tag{3.26}$$

with the average permittivity

$$\widetilde{\varepsilon}_z(n) = \frac{1}{4}\left(\varepsilon(n_{xy})|F_z(n_{xy})| + \varepsilon(n_x)|F_z(n_x)| + \varepsilon(n)|F_z(n)| + \varepsilon(n_y)|F_z(n_y)|\right)$$

and $n_{xy} = n - K_x - K_y$, $n_x = n - K_x$, $n_y = n - K_y$. Note that for the dual facet $\widetilde{F}_z(n)$ it holds

$$\left|\widetilde{F}_z(n)\right| = \frac{1}{4}\left(|F_z(n_{xy})| + |F_z(n_x)| + |F_z(n)| + |F_z(n_y)|\right)$$

and $\varepsilon(n)$ denotes the permittivity on the volume $V(n)$, see Figure 3.6(a). Combining (3.25) and (3.26) yields

$$\widehat{\widehat{d}}_z(n) = \overline{\varepsilon}_z(n)\,\widehat{e}_z(n) + \mathcal{O}\left(h^3\right).$$

with $\overline{\varepsilon}_z(n) = \frac{\widetilde{\varepsilon}_z(n)}{|E_z(n)|}$. Finally we get the *permittivity matrix*

$$M_\varepsilon = \operatorname{diag}\left(\overline{\varepsilon}_x(1), \ldots, \overline{\varepsilon}_x(N), \overline{\varepsilon}_y(1), \ldots, \overline{\varepsilon}_y(N), \overline{\varepsilon}_z(1), \ldots, \overline{\varepsilon}_z(N)\right).$$

The *conductivity matrix* M_σ for linear, inhomogeneous and isotropic conductivities is defined analogously, see [Cle98, Krü00, Ben06]. On a similar way a linear, inhomogeneous and isotropic *reluctivity matrix* can be deduced by taking Figure 3.6(b) into account. For the derivation of a nonlinear, inhomogeneous and anisotropic reluctivity matrix $M_\nu(\widehat{b})$ given by Brauer's model, we refer to [Sch11]. The *discrete constitutive laws* reads:

$$\widehat{\widehat{d}} = M_\varepsilon \widehat{e} \tag{3.27}$$

$$\widehat{\widehat{j}}_c = M_\sigma \widehat{e} \tag{3.28}$$

$$\widehat{h} = M_\nu(\widehat{b})\widehat{b} \tag{3.29}$$

Table 3.6 shows the SI units.

Remark 3.9. The discrete material matrix of permittivities is diagonally positive definite, while the discrete material matrix of conductivities is typically diagonally positive semi-definite if insulators are present in our domain Ω, otherwise positive definite. In case of non-orthogonal grids band structured matrices results. For Brauer's model the discrete material matrix of reluctivities is positive definite.

The basic approximation of the constitutive laws leads to a staircase approximations at curved boundaries. In practice this limitation is overcome by subgridding at boundaries or using other more elaborate schemes, see [TW96, Cle05].

3.2.2 Algebraic Properties of the Discrete Operators

The discrete operators in terms of the FIT have several important inheritances of their continuous counterparts and are composed of simple two-banded matrices, which can be interpreted as discretized partial differential operators, see [BDD$^+$92, CW01b].

Let be $w \in \{x, y, z\}$. We introduce the upper shift matrices $U_w \in \{0, 1\}^{N \times N}$ with

$$(U_w)_{ij} = \delta_{i+K_w, j} \text{ and } U_w = U_x^{K_w}. \tag{3.30}$$

We define the discretized partial differential operators $P_w \in \{-1, 0, 1\}^{N \times N}$ by

$$P_w = U_w - I$$

where P_w is nonsingular, $w \in \{x, y, z\}$. The *discrete curl operators* can be written as

$$C = \begin{bmatrix} 0 & -P_z & P_y \\ P_z & 0 & -P_x \\ -P_y & P_x & 0 \end{bmatrix} \in \{-1, 0, 1\}^{3N \times 3N}$$

and the duality of the two grids yields the simple relation

$$\tilde{C} = C^\mathsf{T}.$$

The *discrete divergence operators* are constructed by

$$S = \begin{bmatrix} P_x & P_y & P_z \end{bmatrix} \in \{-1, 0, 1\}^{N \times 3N} \text{ and } \tilde{S} = \begin{bmatrix} -P_x^\mathsf{T} & -P_y^\mathsf{T} & -P_z^\mathsf{T} \end{bmatrix}.$$

Finally the *discrete gradient operators* are obtained by

$$G = -\tilde{S}^\mathsf{T} \text{ and } \tilde{G} = -S^\mathsf{T},$$

see [CW01b, CW01a].

Lemma 3.10 (Lemma A.1., [Sch11]). Let be $v, w \in \{x, y, z\}$. The relation

$$P_v P_w = P_w P_v$$

holds true.

Proof. Straightforward calculus using (3.30) leads to

$$P_v P_w = (U_v - I)(U_w - I)$$
$$= U_v U_w - U_v - U_w + I$$

67

$$= U_x^{K_v} U_x^{K_w} - U_v - U_w + I$$
$$= U_x^{K_w} U_x^{K_v} - U_v - U_w + I$$
$$= U_w U_v - U_w - U_v + I$$
$$= (U_w - I)(U_v - I)$$
$$= P_w P_v.$$

$\square$

The result reflects the interchange of partial derivatives as in the continuous case, [BDD+92].

Lemma 3.11 ([BDD+92]). The discrete operator identities

$$SC = 0 \qquad\qquad \widetilde{S}\widetilde{C} = 0 \qquad\qquad (3.31)$$
$$CG = 0 \qquad\qquad \widetilde{C}\widetilde{G} = 0$$

hold true.

Proof. To prove the identities we use Lemma 3.10. We get

$$SC = \begin{bmatrix} P_y P_z - P_z P_y & P_z P_x - P_x P_z & P_x P_y - P_y P_x \end{bmatrix} = 0.$$

The dual case is analogous. To show the other identities we simply transpose (3.31). $\square$

That is, the discrete gradient, curl and divergence inherit important operator identities from their continuous counterparts, namely

$$\nabla \cdot \nabla \times \equiv 0 \text{ and } \nabla \times \nabla \equiv 0,$$

which is an important property of the FIT discretization.

We have already seen that the continuity equation can be derived from ME. Due to the properties of the discrete operators given in Lemma 3.11 that is possible in the discrete case, too. From (3.24) we derive the built-in *discrete continuity equation* by

$$\frac{d}{dt}\widetilde{S}\widehat{d} + \widetilde{S}\widehat{j}_c = 0 \qquad\qquad (3.32)$$

which corresponds to the continuous counterpart and is an essential feature of FIT. The discrete continuity equation is of great importance for our later investigations in circuit models including EM devices. The discrete continuity equation ensures that no erroneous charges arises, see [CW01b].

Lemma 3.12. For the discrete operators the relations

$$\ker S = \operatorname{im} C \qquad\qquad \ker \widetilde{S} = \operatorname{im} \widetilde{C}$$
$$\ker C = \operatorname{im} G \qquad\qquad \ker \widetilde{C} = \operatorname{im} \widetilde{G}$$

hold true.

Proof. We apply Lemma 3.10 and 3.11. Exemplarily we show $\ker S = \operatorname{im} C$. Due to Lemma 3.11 we achieve directly $\ker S \supset \operatorname{im} C$. Let be $w \in \ker S$. Then we get

$$P_x w_1 + P_y w_2 + P_z w_3 = 0 \Leftrightarrow w_1 = -P_x^{-1}\left(P_y w_2 + P_z w_3\right).$$

Next we choose $u_1 = 0$, $u_2 = P_x^{-1} w_3$ and $u_3 = -P_x^{-1} w_2$. So we obtain $w = Cu$ and hence $w \in \operatorname{im} C$. The other relations can be deduced in a similar way. □

3.2.3 Discrete Potential Formulation and Gauge Conditions

In analogy to ME we introduce auxiliary functions to simplify the representation of MGE and use a discrete potential approach, see [Cle98, CW99, MMS01].

From discrete Gauss' law for magnetism (3.22) we deduce from a discrete version of Helmholtz decomposition that there is a vector function $\widehat{a} : \mathcal{I} \to \mathbb{R}^{3N}$ such that

$$\widehat{b} = C\widehat{a} \tag{3.33}$$

and using discrete Maxwell-Faraday's law (3.23) we obtain

$$C\left(\overline{e} + \frac{d}{dt}\widehat{a}\right) = 0$$

and conclude, using a discrete version of Helmholtz decomposition, that there is a vector function $\phi : \mathcal{I} \to \mathbb{R}^N$ such that

$$\overline{e} = -G\phi - \frac{d}{dt}\widehat{a}, \tag{3.34}$$

see [Cle05]. This approach is the discrete $(\widehat{a}, \phi)$-formulation with the *discrete vector potential* $\widehat{a}$ and *discrete scalar potential* ϕ. It fulfills immediately the discrete Gauss' law for magnetism and the discrete Maxwell-Faraday's law because important properties from vector calculus are transfered to the discrete level, see Lemma 3.11. A visual representation of all quantities is given in Figure 3.1(b).

As in the continuous case we need a gauge condition to remove the gauge freedom since the discrete curl-operator inherits the non-uniqueness from its continuous counterpart. A common gauge condition approach is the *grad-div regularization*, [CW02], which utilizes the discrete gradient and divergence operator and suitable *discrete artificial material matrices*. This motivates a new *discrete class of gauge conditions* in terms of the FIT given by

$$\vartheta M_\varepsilon G \frac{d}{dt}\phi + M_\zeta GM_\xi \widetilde{S} M_\zeta \widehat{a} = 0 \tag{3.35}$$

where the artificial material matrices M_ζ maps primary edges to dual facets, M_ξ maps dual points to primary volumes and $\vartheta \in \mathbb{R}$ is a "slider" between a type of discrete

Coulomb and Lorenz gauge. The discrete class of gauge conditions (3.35) is the discrete analogon to (3.15). The discrete material matrix M_ξ is called norm matrix and supplies the correct units to the discrete grad-div regularization. In case of $\vartheta = 0$ suitable choices for the discrete material matrices M_ζ and M_ξ are discussed in [CW02, Cle05, BCDS11]. For another type of discrete gauge conditions motivated by damped wave equations we refer to [BCS12].

Assumption 3.13. The discrete artificial material matrices M_ζ and M_ξ are positive definite.

Let Assumption 3.13 be true. For $\vartheta = 0$ we obtain a type of *discrete Coulomb gauge*

$$\widetilde{S} M_\zeta \widehat{a} = 0$$

due to $\widetilde{S} M_\xi G$ is nonsingular. The discrete Coulomb gauge is known from literature. Moreover, for $\vartheta = 1$ we obtain a type of *discrete Lorenz gauge*

$$M_\varepsilon G \frac{d}{dt} \phi + M_\zeta G M_\xi \widetilde{S} M_\zeta \widehat{a} = 0$$

and selecting in addition $M_\zeta = M_\varepsilon$ yields

$$\frac{d}{dt} \phi + M_\xi \widetilde{S} M_\varepsilon \widehat{a} = 0,$$

due to $\widetilde{S} M_\varepsilon G$ is nonsingular. For linear, homogeneous and isotropic materials and an equidistant grid the discrete grad-div regularization regularizes the discrete curl-curl matrix and the resulting discrete operator corresponds to the discrete vector Laplacian.

Lemma 3.14. Let $M \in \mathbb{R}^{N \times N}$ be positive definite. Then, the matrix

$$\overline{C} = \widetilde{C} M_\nu C - G^\top M \widetilde{S}$$

is positive definite.

Proof. We use the relation $G = -\widetilde{S}^\top$. The matrix $\overline{C}$ is symmetric positive semidefinite since $\overline{C}$ is the sum of two positive semidefinite matrices. To show positive definiteness we prove the nonsingularity. Let be $x \in \ker \overline{C}$. Then

$$\left(\widetilde{C} M_\nu C + \widetilde{S}^\top M \widetilde{S} \right) x = 0 \Leftrightarrow Cx = 0 \text{ and } \widetilde{S} x = 0,$$

see Lemma A.3. Hence $x \in \ker C \cap \ker \widetilde{S}$. With Lemma 3.12 it is clear that

$$x \in \ker C \cap \operatorname{im} \widetilde{C} = \ker C \cap (\ker C)^\perp$$

and hence $x = 0$. $\qquad\square$

Note that $\widetilde{C} M_\nu C + M_\zeta \widetilde{S}^\top M_\xi \widetilde{S} M_\zeta$ is not necessarily positive definite. Hence not an arbitrary type of discrete Coulomb or Lorenz gauge leads to a gauge condition in the sense of a discrete curl-curl operator regularization.

3.2.4 Phantom Objects and Discrete Boundary Conditions

The numbering scheme of the grid $\mathcal{G}$ introduces needless phantom objects at the boundary. The *phantom objects* are edges, facets and volumes which have to be disregarded. In order to disregard those objects we follow and extend the idea of [Doh92, Sch11]. Table 3.7 gives an overview of the number of non-phantom objects.

object	number of non-phantom objects
primary nodes/dual volumes	$N_x N_y N_z$
primary facets/dual volumes	$(N_x - 1) N_y N_z$ $+ N_x (N_y - 1) N_z$ $+ N_x N_y (N_z - 1)$
primary facets/dual edges	$N_x (N_y - 1)(N_z - 1)$ $+ (N_x - 1) N_y (N_z - 1)$ $+ (N_x - 1)(N_y - 1) N_z$
primary volumes/dual nodes	$(N_x - 1)(N_y - 1)(N_z - 1)$

Table 3.7: Number of non-phantom objects.

Example 3.15. Regarding the primary FIT grid of two points in each direction as shown in Figure 3.7. The grid consists of 8 nodes, 12 edges, 6 facets and one volume. The numbering scheme introduces 12 edges, 18 facets and 7 volumes which are needless.

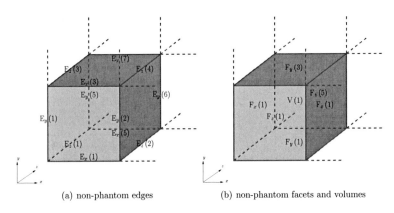

(a) non-phantom edges   (b) non-phantom facets and volumes

Figure 3.7: Primary FIT grid of dimensions $2 \times 2 \times 2$ with non-phantom objects.

To find all phantom objects it is sufficient to characterize the phantom edges. These edges are always attached to points on the boundary that are addressed by $n(i_x, i_y, i_z)$,

71

where one direction reaches its maximum $i_w = N_w$ with $w \in \{x, y, z\}$. For each $w \in \{x, y, z\}$ the set

$$\mathcal{H}_w = \{1 \leqslant n\,(i_x, i_y, i_z) \leqslant N | i_w = N_w\}$$

contains the indices of all points with an attached phantom edge in direction w and we fade-out all phantom objects using the diagonal fade-out matrix $F_w \in \{0, 1\}^{N \times N}$ given by:

$$(F_w)_{ij} = \begin{cases} 1, & \text{for } i = j \text{ and } i \notin \mathcal{H}_w, \\ 0, & \text{else.} \end{cases}$$

The sets $\mathcal{H}_{xyz} = \mathcal{H}_x \cup \mathcal{H}_y \cup \mathcal{H}_z$ and $\mathcal{H}_{vw} = \mathcal{H}_v \cup \mathcal{H}_w$ contain all points connected to at least one phantom edge and connected to at least one phantom edge in the $(v\text{-}w)$-plane with $v, w \in \{x, y, z\}$, $v \neq w$, respectively. Next we investigate some properties of the fade-out matrices.

Lemma 3.16 (Lemma A.3., [Sch11]). The matrices F_w are orthogonal projectors and for $v \neq w$

$$F_w F_v = F_v F_w, \tag{3.36}$$

$$F_w F_v P_v = F_v P_v F_w \tag{3.37}$$

is valid with $v, w \in \{x, y, z\}$.

Proof. The projector properties of F_w as well as (3.36) are clear since they are diagonal matrices containing only zeros and ones. The left-hand side of (3.37) reads:

$$(F_w F_v P_v)_{ij} = \begin{cases} -1, & \text{for } j = i \text{ and } i \notin \mathcal{H}_{wv}, \\ 1, & \text{for } j = i + K_v \text{ and } i \notin \mathcal{H}_{wv}, \\ 0, & \text{else.} \end{cases}$$

The right-hand side of (3.37) reads:

$$(F_v P_v F_w)_{ij} = \begin{cases} -1, & \text{for } j = i \text{ and } i \notin \mathcal{H}_{wv}, \\ 1, & \text{for } j = i + K_v, \ j \notin \mathcal{H}_w \text{ and } i \notin \mathcal{H}_v, \\ 0, & \text{else.} \end{cases}$$

Now we show that both sides equals. Since $i \notin \mathcal{H}_v$ we can write

$$i = i_x + i_y K_y + i_z K_z, \text{ with } i_v < N_v$$

and thus $j = i + K_v$ gives

$$j = j_x + j_y K_y + j_z K_z, \text{ with } j_v = i_v + 1 \leqslant N_v.$$

Then we know that $j_w = i_w$ since $v \neq w$ and thus the condition $i \notin \mathcal{H}_w$ is equivalent to $j = i + K_v \notin \mathcal{H}_w$ for $v \neq w$. $\qquad \square$

All points addressed by the numbering scheme are included in the primary grid but only subsets of the addressed edges, facets and volumes really exist. Only edges not in $\mathcal{H}_w$, $w \in \{x, y, z\}$, exists. Furthermore facets and volumes exists if and only if all their edges exist. To fade-out the phantom objects we define

$$F_N = I, \; F_E = \begin{bmatrix} F_x & 0 & 0 \\ 0 & F_y & 0 \\ 0 & 0 & F_z \end{bmatrix} F_F, = \begin{bmatrix} F_y F_z & 0 & 0 \\ 0 & F_x F_z & 0 \\ 0 & 0 & F_x F_y \end{bmatrix} \text{ and } F_V = F_x F_y F_z$$

where F_N, F_E, F_F and F_V denote the *fade-out projectors* for all points, edges, facets and volumes in the primary grid. Analogously we define the corresponding counterparts for the dual grid and benefit of the relation between both grids.

Corollary 3.17. The fade-out projectors F_N, F_E, F_F and F_V are orthogonal projectors.

$\square$

Next we define the discrete operators with fade-out phantom objects. The gradient operator maps points to edges and we have to ignore contributions from phantom points and edges. We achieve

$$G_F = F_E G F_N \text{ and } \tilde{G}_F = F_F \tilde{G} F_V.$$

The curl operator maps edges to facets and therefore we have to ignore contributions from phantom edges and facets. We get

$$C_F = F_F C F_E \text{ and } \tilde{C}_F = F_E \tilde{C} F_F.$$

In the end the divergence operator maps facets to volumes and hence we have to ignore contributions from phantom facets and volumes. We gain

$$S_F = F_V S F_F \text{ and } \tilde{S}_F = F_N \tilde{S} F_E.$$

All discrete operators with fade-out phantom objects have a redundancy.

Lemma 3.18 (Corollary A.5., [Sch11]). For the discrete operator with fade-out phantom objects the relations

$$G_F = F_E G \qquad\qquad \tilde{G}_F = \tilde{G} F_V$$
$$C_F = F_F C \qquad\qquad \tilde{C}_F = \tilde{C} F_F$$
$$S_F = F_V S \qquad\qquad \tilde{S}_F = \tilde{S} F_E$$

hold true.

Proof. This is a consequence of Lemma 3.16. $\square$

For the discrete operators with fade-out phantom objects all important properties still hold true.

Lemma 3.19 (Theorem A.6., [Sch11]). The discrete operator identities

$$S_F C_F = 0 \qquad\qquad \tilde{S}_F \tilde{C}_F = 0$$
$$C_F G_F = 0 \qquad\qquad \tilde{C}_F \tilde{G}_F = 0$$

hold true.

Proof. This is a consequence of Lemma 3.11 and 3.18. $\qquad\square$

Remark 3.20. Sometimes in literature the discrete partial differential operators are directly constructed as $\overline{P}_w = F_w P_w$ with $w \in \{x, y, z\}$, for example, as in [Ben06].

Not only in the discrete operators the phantom objects occur but also in the discrete material matrices. The discrete permittivity matrix as well as the discrete conductivity matrix are mapped from primary edges to dual facets and we get

$$M_\varepsilon^F = F_E M_\varepsilon F_E \text{ and } M_\sigma^F = F_E M_\sigma F_E.$$

Furthermore, the discrete reluctivity matrix maps from primary facets to dual edges and reads

$$M_\nu^F = F_F M_\mu F_F.$$

In fact we do not simply want to fade-out the phantom objects but we want to delete the corresponding rows and columns within the discrete operators, too. For that we extend the idea of phantom objects of [Sch11] and we define the matrices

$$D_w \in \{0,1\}^{N-|\mathcal{H}_w| \times N}, \ D_{vw} \in \{0,1\}^{N-|\mathcal{H}_{vw}| \times N}, \ D_{xyz} \in \{0,1\}^{N-|\mathcal{H}_{xyz}| \times N}$$

related to the fade-out projectors by

$$D_w^T D_w = F_w \qquad\qquad D_w D_w^T = I$$
$$D_{vw}^T D_{vw} = F_v F_w \qquad\qquad D_{vw} D_{vw}^T = I$$
$$D_{xyz}^T D_{xyz} = F_x F_y F_z \qquad\qquad D_{xyz} D_{xyz}^T = I$$

with $v, w \in \{x, y, z\}$, $v \neq w$. We construct the *deletion* and *shrinking* matrices

$$D_N = I, \ D_E = \begin{bmatrix} D_x & 0 & 0 \\ 0 & D_y & 0 \\ 0 & 0 & D_z \end{bmatrix}, \ D_F = \begin{bmatrix} D_{yz} & 0 & 0 \\ 0 & D_{xz} & 0 \\ 0 & 0 & D_{xy} \end{bmatrix} \text{ and } D_V = D_{xyz}$$

where D_N, D_E, D_F and D_V denotes the matrix deleting all rows of F_N, F_E, F_F and F_V belonging to phantom objects. For the deletion matrices the relations

$$D_N^T D_N = F_N \qquad\qquad D_N D_N^T = I$$
$$D_E^T D_E = F_E \qquad\qquad D_E D_E^T = I$$

$$D_F^\mathsf{T} D_F = F_F \qquad\qquad D_F D_F^\mathsf{T} = I$$
$$D_V^\mathsf{T} D_V = F_V \qquad\qquad D_V D_V^\mathsf{T} = I$$

hold true.

Now we remove the phantom objects. For that we left-multiply the different discretized laws of MGE by the corresponding deletion matrix and we set the unknowns corresponding to phantom objects to zero. From MGE (3.21) to (3.24) we deduce the phantom-free MGE

$$\tilde{S}_D \widehat{\widehat{d}}_D = q_D \tag{3.38}$$

$$S_D \widehat{b}_D = 0 \tag{3.39}$$

$$C_D \widehat{e}_D = -\frac{d}{dt} \widehat{b}_D \tag{3.40}$$

$$\tilde{C}_D \widehat{h}_D = \frac{d}{dt} \widehat{\widehat{d}}_D + \widehat{\widehat{j}}_{c,D} \tag{3.41}$$

where the existing unknowns are

$$\widehat{e}_D = D_E \widehat{e}, \ \ \widehat{\widehat{d}}_D = D_E \widehat{\widehat{d}}, \ \ \widehat{\widehat{j}}_{c,D} = D_E \widehat{\widehat{j}}_c, \ \ \widehat{h}_D = D_F \widehat{h}, \ \ \widehat{b}_D = D_F \widehat{b} \text{ and } q_D = D_N q$$

and for the phantom-free operators the relations

$$\tilde{C}_D = D_E \tilde{C} D_F^\mathsf{T}, \ \ \tilde{S}_D = D_N \tilde{S} D_E^\mathsf{T}, \ \ S_D = D_V S D_F^\mathsf{T} \text{ and } C_D = D_F C D_E^\mathsf{T}$$

hold true. Applying the same deduction as above, the discrete constitutive laws (3.27), (3.28) and (3.29) yields

$$\widehat{\widehat{d}}_D = M_\varepsilon^D \widehat{e}_D \tag{3.42}$$

$$\widehat{\widehat{j}}_{c,D} = M_\sigma^D \widehat{e}_D \tag{3.43}$$

$$\widehat{h}_D = M_\nu^u(\widehat{b}_D) \widehat{b}_D \tag{3.44}$$

where the phantom-free material matrices are given by

$$M_\varepsilon^D = D_E M_\varepsilon D_E^\mathsf{T}, \ M_\sigma^D = D_E M_\sigma D_E^\mathsf{T} \text{ and } M_\nu^D = D_F M_\nu D_F^\mathsf{T}.$$

The discrete equations for the vector and scalar potential (3.33) and (3.34) result in

$$\widehat{b}_D = C_D \widehat{a}_D \tag{3.45}$$

$$\widehat{e}_D = -G_D \phi_D - \frac{d}{dt} \widehat{a}_D \tag{3.46}$$

at which the existing unknowns are

$$\phi_D = D_N \phi \text{ and } \widehat{a}_D = D_E \widehat{a}$$

and the phantom-free gradient reads $G_D = D_E G D_N^\mathsf{T}$.

Remark 3.21. By $\mathcal{E}_F \subset \mathcal{E}$ we denote the set of non-phantom edges indices. We can interpret G_D as the transpose incidence matrix of the directed graph $(\mathcal{N}, \mathcal{E}_F)$. We will come back to that later when motivating the boundary conditions.

For completeness we have $\widetilde{G}_D = D_F \widetilde{G} D_V^T$. The phantom-free operators still have the following properties:

Lemma 3.22. The phantom-free operators identities

$$S_D C_D = 0 \qquad\qquad \widetilde{S}_D \widetilde{C}_D = 0$$
$$C_D G_D = 0 \qquad\qquad \widetilde{C}_D \widetilde{G}_D = 0$$

hold true.

Proof. This is a consequence of Lemma 3.19. □

Using (3.41) and Lemma 3.22 we can derive the phantom-free continuity equation given by

$$\frac{d}{dt} \widetilde{S}_D \widehat{d}_D + \widetilde{S}_D \widehat{j}_{c,D} = 0 \tag{3.47}$$

but note that we can derive (3.47) also directly from (3.32). Therefore charge conservation is also fulfilled for the phantom-free operators.

From the phantom-free equation for the vector potential (3.45) and scalar potential (3.46) we can deduce that the phantom-free Gauss' law for magnetism (3.39) and Maxwell-Faraday's law (3.40) are fulfilled automatically like in the continuous case. Based on (3.47) we conclude that if

$$\widetilde{S}_D \widehat{d}_D (t_0) = q_D (t_0), \ t_0 \in \mathcal{I},$$

then the phantom-free Gauss' law (3.38) is always fulfilled like in the continuous case. That is, it is sufficient to take the phantom-free Maxwell-Ampère's law (3.41) into account using the phantom-free potential formulation.

Boundary Conditions

The next step is to incorporate the boundary conditions. Here we focus on the PEC case and apply Dirichlet boundary conditions for the unknowns.

Let $\Omega_N = \{1, \ldots, N\}$ be the set of node indices and $\Gamma_N = \{n \in \Omega_N | N(n) \in \Gamma\}$ the set of boundary node indices. We denote by $\Omega_N^C = \Omega_N \backslash \Gamma_N$ the set of non-boundary node indices and $n_\phi = |\Omega_N^C|$. To describe the non-phantom edge indices properly we need some notation. Let

$$\mathcal{E}_x = \{n \in \mathbb{N} | n \in \mathcal{H}_x\} \qquad \mathcal{E}_y = \{n + N \in \mathbb{N} | n \in \mathcal{H}_y\} \qquad \mathcal{E}_z = \{n + 2N \in \mathbb{N} | n \in \mathcal{H}_z\}$$

and

$$\overline{\mathcal{E}}_x = \{1,\ldots,N\}\setminus\mathcal{E}_x \qquad \overline{\mathcal{E}}_y = \{N+1,\ldots,2N\}\setminus\mathcal{E}_y \qquad \overline{\mathcal{E}}_z = \{2N+1,\ldots,3N\}\setminus\mathcal{E}_z$$

be the index sets of phantom and non-phantom edges in each direction. Then the set of non-phantom edges indices is given by $\mathcal{E}_F = \overline{\mathcal{E}}_x \cup \overline{\mathcal{E}}_y \cup \overline{\mathcal{E}}_z$. Let $\Omega_E = \{1,\ldots,|\mathcal{E}_F|\}$ be the set of the renumbered non-phantom edge indices. For renumbering the non-phantom edges we define the injective and surjective mapping

$$\mathrm{p}:\{1,\ldots,3N\}\to\Omega_E$$

with the property

$$i<j\Leftrightarrow\mathrm{p}(i)<\mathrm{p}(j),\ i,j\in\{1,\ldots,3N\},$$

where for the preimage

$$\mathrm{p}^{-1}(k)\in\mathcal{E}_F,\ k\in\Omega_E$$

holds true. By

$$\Gamma_E = \left\{n\in\Omega_E \mid m=\mathrm{p}^{-1}(n),m\in\overline{\mathcal{E}}_w \text{ and } \mathrm{E}_m(w)\subset\Gamma, w\in\{x,y,z\}\right\}$$

we denote the set of renumbered boundary edge indices and $\Omega_E^C = \Omega_E\setminus\Gamma_E$ is the set of renumbered non-boundary non-phantom edge indices, where the non-boundary non-phantom edges are degrees of freedom. We denote $n_a = |\Omega_E^C|$.

We introduce the *unknown* and *boundary selection matrices*

$$\mathrm{U}_N\in\{0,1\}^{n_\phi\times N},\ \mathrm{U}_E\in\{0,1\}^{n_a\times|\mathcal{E}_F|},\ \mathrm{B}_N\in\{0,1\}^{|\Gamma_N|\times N},\ \mathrm{B}_E\in\{0,1\}^{|\Gamma_E|\times|\mathcal{E}_F|}$$

for nodes and edges defined by

$$\mathrm{U}_N^\mathsf{T}\mathrm{U}_N=\mathrm{U}_{F,N} \qquad \mathrm{U}_N\mathrm{U}_N^\mathsf{T}=\mathrm{I} \qquad \mathrm{B}_N^\mathsf{T}\mathrm{B}_N=\mathrm{B}_{F,N} \qquad \mathrm{B}_N\mathrm{B}_N^\mathsf{T}=\mathrm{I}$$
$$\mathrm{U}_E^\mathsf{T}\mathrm{U}_E=\mathrm{U}_{F,E} \qquad \mathrm{U}_E\mathrm{U}_E^\mathsf{T}=\mathrm{I} \qquad \mathrm{B}_E^\mathsf{T}\mathrm{B}_E=\mathrm{B}_{F,E} \qquad \mathrm{B}_E\mathrm{B}_E^\mathsf{T}=\mathrm{I}$$

with the properties

$$\mathrm{U}_{F,N}+\mathrm{B}_{F,N}=\mathrm{F}_N \text{ and } \mathrm{U}_{F,E}+\mathrm{B}_{F,E}=\mathrm{F}_E.$$

With that we obtain the relations

$$\mathrm{U}_N=\mathrm{U}_N\mathrm{U}_{F,N} \qquad \mathrm{U}_N^\mathsf{T}=\mathrm{U}_{F,N}\mathrm{U}_N^\mathsf{T} \qquad \mathrm{B}_N=\mathrm{B}_N\mathrm{B}_{F,N} \qquad \mathrm{B}_N^\mathsf{T}=\mathrm{B}_{F,N}\mathrm{B}_N^\mathsf{T} \qquad \mathrm{U}_N\mathrm{B}_{F,N}=0$$

and

$$\mathrm{U}_E=\mathrm{U}_E\mathrm{U}_{F,E} \qquad \mathrm{U}_E^\mathsf{T}=\mathrm{U}_{F,E}\mathrm{U}_E^\mathsf{T} \qquad \mathrm{B}_E=\mathrm{B}_E\mathrm{B}_{F,E} \qquad \mathrm{B}_E^\mathsf{T}=\mathrm{B}_{F,E}\mathrm{B}_E^\mathsf{T} \qquad \mathrm{U}_E\mathrm{B}_{F,E}=0$$

hold true

At this point we need the orthogonality of each grid complex $\mathcal{G}$ and $\widetilde{\mathcal{G}}$.

Remark 3.23. For every diagonal matrix $D \in \mathbb{R}^{|\mathcal{E}_F| \times |\mathcal{E}_F|}$ we get $U_E D B_E^\top = 0.$ $\qquad\qquad\square$

Now we incorporate the boundary conditions into the equations. We start with the phantom-free constitutive laws. Left-multiplying (3.42) by $U_E D_E^\top$ and using Remark 3.23 we acquire on the one hand

$$\begin{aligned}
U_E D_E^\top \widehat{\widehat{d}}_D &= U_E D_E^\top M_\varepsilon^D \widehat{e}_D \\
&= U_E D_E^\top D_E M_\varepsilon D_E^\top D_E \widehat{e} \\
&= U_E F_E M_\varepsilon F_E \widehat{e} \\
&= U_E M_\varepsilon \left(U_{F,E} + B_{F,E} \right) \widehat{e} \\
&= U_E M_\varepsilon U_{F,E} \widehat{e} \\
&= U_E M_\varepsilon U_E^\top U_E \widehat{e}
\end{aligned}$$

and on the other hand

$$\begin{aligned}
U_E D_E^\top \widehat{\widehat{d}}_D &= U_E D_E^\top D_E \widehat{\widehat{d}} \\
&= U_E F_E \widehat{\widehat{d}} \\
&= U_E \widehat{\widehat{d}}.
\end{aligned}$$

With that the phantom-free constitutive laws (3.42), (3.43) and (3.44) yields the *reduced discrete constitutive laws*

$$\widehat{\widehat{d}}_u = M_\varepsilon^u \widehat{e}_u, \tag{3.48}$$

$$\widehat{\widehat{j}}_{c,u} = M_\sigma^u \widehat{e}_u, \tag{3.49}$$

$$\widehat{h}_u = M_\nu^u(\widehat{\widehat{b}}_u)\widehat{\widehat{b}}_u \tag{3.50}$$

with the unknowns

$$\widehat{e}_u = U_E \widehat{e}, \ \ \widehat{\widehat{b}}_u = \widehat{\widehat{b}}_D, \ \ \widehat{h}_u = \widehat{h}_D, \ \ \widehat{\widehat{d}}_u = U_E \widehat{\widehat{d}} \ \text{ and } \ \widehat{\widehat{j}}_{c,u} = U_E \widehat{\widehat{j}}_c$$

and reduced discrete material matrices

$$M_\varepsilon^u = U_E M_\varepsilon U_E^\top, \ \ M_\sigma^u = U_E M_\sigma U_E^\top \ \text{ and } \ M_\nu^u(\widehat{\widehat{b}}_u) = M_\nu^D(\widehat{\widehat{b}}_D).$$

From phantom-free Maxwell-Ampère's law (3.41) we get the *reduced discrete Maxwell-Ampère's law*

$$\widetilde{C}_u \widehat{h}_u = \frac{d}{dt} \widehat{\widehat{d}}_u + \widehat{\widehat{j}}_{c,u} \tag{3.51}$$

with the *reduced discrete dual curl operator*

$$\widetilde{C}_u = U_E \widetilde{C} D_F^\top.$$

From the equations of the phantom-free vector and scalar potential (3.45) and (3.46) we obtain

$$\widehat{b}_u = C_u \widehat{a}_u + C_b \widehat{a}_b \tag{3.52}$$

$$\widehat{e}_u = -G_u \phi_u - G_b \phi_b - \frac{d}{dt} \widehat{a}_u \tag{3.53}$$

at which the unknowns read

$$\phi_u = U_N \phi, \quad \phi_b = B_N \phi, \quad \widehat{a}_u = U_E \widehat{a} \text{ and } \widehat{a}_u = B_E \widehat{a}$$

and reduced discrete operators are given by

$$G_u = U_E G U_N^T, \quad G_b = U_E G B_N^T, \quad C_u = D_F C U_E^T \text{ and } C_b = D_F C B_E^T.$$

In addition the *reduced discrete dual divergence operator* reads

$$\widetilde{S}_u = U_N \widetilde{S} U_E^T$$

We show that important reduced discrete operator identities are still valid.

Lemma 3.24. The relations $G_u = -\widetilde{S}_u^T$ and $\widetilde{C}_u = C_u^T$ hold true. □

Lemma 3.25. The reduced discrete operator identity $\widetilde{S}_u \widetilde{C}_u = 0$ and $C_u G_u = 0$ hold true.

Proof. We infer from $G_u = U_E G U_N^T$ that $U_E^T G_u U_N = U_{F,E} G U_{F,N}$. For every edge in Γ_E the front and back node are in Γ_N. That is, $U_{F,N}$ set exactly that columns to zero which are not effected by $U_{F,E}$ Hence $G U_{F,N} = U_{F,E} G U_{F,N}$ and $G U_N^T = U_{F,E} G U_N^T$. We get

$$\begin{aligned}
\widetilde{S}_u \widetilde{C}_u &= U_N \widetilde{S} U_E^T U_E \widetilde{C} D_F^T \\
&= U_N \widetilde{S} U_{F,E} \widetilde{C} D_F^T \\
&= U_N \widetilde{S} \widetilde{C} D_F^T \\
&= 0,
\end{aligned}$$

see Lemma 3.11. The other statement follows directly. □

Using (3.51) and Lemma 3.25 we can derive the *reduced discrete continuity equation* given by

$$\frac{d}{dt} \widetilde{S}_u \widehat{\widehat{d}}_u + \widetilde{S}_u \widehat{\widehat{j}}_{c,u} = 0 \tag{3.54}$$

but note that we can (3.54) derive also directly from (3.47). Left-multiplying (3.47) by $U_N D_N^T$ we obtain

$$0 = \frac{d}{dt} U_N D_N^T \widetilde{S}_D \widehat{\widehat{d}}_D + U_N D_N^T \widetilde{S}_D \widehat{\widehat{j}}_{c,D}$$

$$= \frac{d}{dt} U_N D_N^T D_N \tilde{S} D_E^T D_E \widehat{\widehat{d}} + U_N D_N^T D_N \tilde{S} D_E^T D_E \widehat{\widehat{j}}_c$$

$$= \frac{d}{dt} U_N F_N \tilde{S} F_E \widehat{\widehat{d}} + U_N F_N \tilde{S} F_E \widehat{\widehat{j}}_c$$

$$= \frac{d}{dt} U_N \tilde{S} \left(U_{F,E} + B_{F,E} \right) \widehat{\widehat{d}} + U_N \tilde{S} \left(U_{F,E} + B_{F,E} \right) \widehat{\widehat{j}}_c$$

$$= \frac{d}{dt} U_N \tilde{S} U_{F,E} \widehat{\widehat{d}} + U_N \tilde{S} U_{F,E} \widehat{\widehat{j}}_c$$

$$= \frac{d}{dt} U_N \tilde{S} U_E^T U_E \widehat{\widehat{d}} + U_N \tilde{S} U_E^T U_E \widehat{\widehat{j}}_c$$

$$= \frac{d}{dt} \tilde{S}_u \widehat{\widehat{d}}_u + \tilde{S}_u \widehat{\widehat{j}}_{c,u}$$

using $B_{F,E} G U_N^T = 0$ since $U_{F,N}$ sets exactly that columns to zero which are effected by $B_{F,E}$. Therefore charge conservation is also fulfilled.

We already mentioned that G_D can be interpreted as the transpose incidence matrix of the directed graph $(\mathcal{N}, \mathcal{E}_F)$, see Remark 3.21. In fact G_u^T is a kind of reduced incidence matrix of the directed graph $(\mathcal{N}, \mathcal{E}_F)$ with more then one reference node, due to setting Dirichlet boundary conditions for all boundary nodes and edges. That is an important observation for the later index analysis of the resulting DAE from MGE.

Remark 3.26. The reduced discrete operator G_u has full column rank. □

3.2.5 Maxwell's Grid Equations with Boundary Excitation

In this subsection we formulate a new *class of reduced discrete gauge conditions* in terms of FIT for the non-phantom and non-boundary unknowns. We describe the boundary conditions for the scalar potential as excitation of the EM fields at the boundary and formulate the current through the EM devices. The excitation and current formulation play a vital role for the circuit models including EM devices.

Motivated by (3.35) we reformulate the class of discrete gauge conditions into the class of reduced discrete gauge conditions given by

$$\vartheta M_\varepsilon^u G_u \frac{d}{dt} \phi_u + M_\nu^u \widehat{a}_u = 0 \qquad (3.55)$$

with *reduced discrete artificial material matrices* given by

$$M_\nu^u = M_\zeta^u G_u M_\xi^u \tilde{S}_u M_\zeta^u, \ M_\zeta^u = U_E M_\zeta U_E^T \text{ and } M_\xi^u = U_N M_\xi U_N^T.$$

Note, we cannot deduce (3.55) directly from (3.35) due to the presence of the boundary conditons but (3.55) is motivated by that. For later investigation we left-multiply (3.55) by $\tilde{S}_u$ and we regard

$$\vartheta \tilde{S}_u M_\varepsilon^u G_u \frac{d}{dt} \phi_u + \tilde{S}_u M_\nu^u \widehat{a}_u = 0. \qquad (3.56)$$

The next step is to describe the Dirichlet boundary conditions for the discrete boundary scalar potentials given by $\phi_b : \mathcal{I} \to \mathbb{R}^{|\Gamma_N|}$ in more detail to have an excitation for the EM fields at the domain boundary. Here we do not follow the approach presented in [DHW04, Ben06, BBS11, Sch11], where the excitation is constructed using different conductor models and applied as a source term. In Subsection 3.1.2 we motivate spatial-constant time-dependent Dirichlet boundary conditions for scalar potentials on each $\Gamma_{N,i}$ with

$$\Gamma_{N,i} = \{n \in \Omega_N | N(n) \in \Gamma_i\}.$$

Without loss of generality we suppose that the first $n_E < k$ boundary parts $\Gamma_{N,i}$ are conductive contacts and $\Gamma_{N,g} = \bigcup_{i=n_E+1}^{k} \Gamma_{N,i}$ is the mass contact. That is, the EM device has $n_E + 1$ contacts. At the mass contact we apply zero potential. The potentials at the conductive contacts are described by $v_E : \mathcal{I} \to \mathbb{R}^{n_E}$. Next we construct an *pre-excitation matrix* $X \in \mathbb{R}^{N \times n_E}$ defined by

$$(X)_{ij} = \begin{cases} 1, & \text{if } i \in \Gamma_{N,j}, \\ 0, & \text{else.} \end{cases}$$

which maps from conductive contacts to nodes, acting only on boundary nodes at conductive contacts. Note that we directly skip the mass contact because of the zero potential. We write the boundary conditions in terms of the input function and obtain the *boundary excitation*

$$\phi_b = B_N X v_E.$$

With the pre-excitation matrix we define *excitation matrix* $\Lambda_u \in \mathbb{R}^{n_a \times n_E}$ by

$$\Lambda_u = -G_b B_N X \tag{3.57}$$

which maps from conductive contacts to non-phantom and non-boundary edges. Due to that construction it is obvious that the excitation matrix acts only on edges attached to conductive contacts and Λ_u has full column rank.

Assumption 3.27. We assume homogeneous Dirichlet boundary condition for the discrete vector potential, that is, $\widehat{a}_b = 0$, and that the applied potential at the mass contact is zero.

Let Assumption 3.27 be true. The applied potential at the EM device conductive contacts generates currents and the *reduced discrete total current density* in terms of FIT is given by reduced discrete Maxwell-Ampère's law (3.51). We get

$$\begin{aligned} \widehat{}\!\!\!\widehat{j}_{t,u} &= \frac{d}{dt} \widehat{}\!\!\!\widehat{d}_u + \widehat{}\!\!\!\widehat{j}_{c,u} \\ &= \widetilde{C}_u M_\nu^u (C_u \widehat{a}_u) C_u \widehat{a}_u. \end{aligned}$$

For our later investigation we are interested in the reduced discrete total current density at the conductive contacts. Here we can utilize the excitation matrix Λ_u. The reduced discrete total current density at a contact is the sum of all contributions from non-phantom and non-boundary edges attached to the contact taking the edges orientation into account. That is, the reduced discrete total current density at the conductive contacts can be described by

$$\mathbf{j}_E = \Lambda_u^\top \widetilde{C}_u M_\nu^u (C_u \widehat{a}_u) C_u \widehat{a}_u \in \mathbb{R}^{n_E}. \tag{3.58}$$

Assumption 3.28. For a consistent contact formulation we assume:

(i) There are at least two conductive contacts.

(ii) The conductive contacts are disjoint and simply connected.

(iii) Between two conductive contacts are at least two primary surfaces.

To ensure this we need a sufficiently fine spatial discretization of the EM device.

To show that we have a consistent discrete contact current formulation we formulate the following lemma.

Lemma 3.29. Let Assumption 3.28 be true. The matrix $C_u \Lambda_u$ has full column rank.

Proof. Regarding the j-th column of $C_u \Lambda_u$. The i-th row of the j-th column of $C_u \Lambda_u$ is nonzero if the i-th primary facets consists of one or three primary edges connected with the j-th conductive contact. Such a primary facet always exists for each conductive contact. If the i-th row of the j-th column is nonzero the other columns are zero at the i-th row. $\qquad \square$

Next we derivative the *reduced curl-curl equation*. Starting with the reduced discrete Maxwell-Ampère's law (3.51), the reduced discrete constitutive laws (3.48), (3.49) and (3.50), using the excitation matrix (3.57) and formulated in terms of the reduced potentials (3.52) and (3.53) we gain

$$
\begin{aligned}
0 &= \widetilde{C}_u \widehat{h}_u - \frac{d}{dt} \widehat{\widehat{d}}_u - \widehat{\widehat{j}}_{c,u} \\
&= \widetilde{C}_u M_\nu^u (\widehat{b}_u) \widehat{b}_u - M_\varepsilon^u \frac{d}{dt} \widehat{e}_u - M_\sigma^u \widehat{e}_u \\
&= \widetilde{C}_u M_\nu^u (C_u \widehat{a}_u) C_u \widehat{a}_u + M_\varepsilon^u \frac{d}{dt} \left(G_u \phi_u + G_b \phi_b + \frac{d}{dt} \widehat{a}_u \right) + M_\sigma^u \left(G_u \phi_u + G_b \phi_b + \frac{d}{dt} \widehat{a}_u \right)
\end{aligned}
$$

and we are ending with the reduced curl-curl equation

$$
\begin{aligned}
0 = M_\varepsilon^u G_u \frac{d}{dt} \phi_u + M_\varepsilon^u \frac{d^2}{dt^2} \widehat{a}_u + M_\sigma^u G_u \phi_u + \widetilde{C}_u M_\nu^u (C_u \widehat{a}_u) C_u \widehat{a}_u + M_\sigma^u \frac{d}{dt} \widehat{a}_u \\
- M_\varepsilon^u \Lambda_u \frac{d}{dt} v_E - M_\sigma^u \Lambda_u v_E.
\end{aligned} \tag{3.59}
$$

Grouping a reduced discrete gauge conditions given by (3.56), the reduced discrete total current density at the conductive contacts (3.58) and the reduced curl-curl equation (3.59) we obtain the MGE

$$\mathrm{j_E} - \Lambda_\mathrm{u}^\top \widetilde{\mathrm{C}}_\mathrm{u} \mathrm{M}_\nu^\mathrm{u} (\mathrm{C}_\mathrm{u} \widehat{\mathrm{a}}_\mathrm{u}) \mathrm{C}_\mathrm{u} \widehat{\mathrm{a}}_\mathrm{u} = 0$$

$$\vartheta \widetilde{\mathrm{S}}_\mathrm{u} \mathrm{M}_\varepsilon^\mathrm{u} \mathrm{G}_\mathrm{u} \frac{\mathrm{d}}{\mathrm{dt}} \phi_\mathrm{u} + \widetilde{\mathrm{S}}_\mathrm{u} \mathrm{M}_\nu^\mathrm{u} \widehat{\mathrm{a}}_\mathrm{u} = 0$$

$$\mathrm{M}_\varepsilon^\mathrm{u} \mathrm{G}_\mathrm{u} \frac{\mathrm{d}}{\mathrm{dt}} \phi_\mathrm{u} + \mathrm{M}_\varepsilon^\mathrm{u} \frac{\mathrm{d}}{\mathrm{dt}} \widehat{\pi}_\mathrm{u} + \mathrm{M}_\sigma^\mathrm{u} \mathrm{G}_\mathrm{u} \phi_\mathrm{u} + \widetilde{\mathrm{C}}_\mathrm{u} \mathrm{M}_\nu^\mathrm{u} (\mathrm{C}_\mathrm{u} \widehat{\mathrm{a}}_\mathrm{u}) \mathrm{C}_\mathrm{u} \widehat{\mathrm{a}}_\mathrm{u} + \mathrm{M}_\sigma^\mathrm{u} \widehat{\pi}_\mathrm{u} \qquad (3.60)$$

$$-\mathrm{M}_\varepsilon^\mathrm{u} \Lambda_\mathrm{u} \frac{\mathrm{d}}{\mathrm{dt}} \mathrm{v_E} - \mathrm{M}_\sigma^\mathrm{u} \Lambda_\mathrm{u} \mathrm{v_E} = 0$$

$$\frac{\mathrm{d}}{\mathrm{dt}} \widehat{\mathrm{a}}_\mathrm{u} - \widehat{\pi}_\mathrm{u} = 0$$

with the *auxiliary vector* $\widehat{\pi}_\mathrm{u}$ to avoid the second-order differentiation in time for $\widehat{\mathrm{a}}_\mathrm{u}$. The number of non-boundary nodes is n_ϕ, the number of non-boundary edges is n_a, n_π and the number of conductive contacts by n_E. The given vector function $\mathrm{v_E}(\mathrm{t})$ describes the applied potential at the conductive contacts in time t, $\mathcal{I} = [\mathrm{t}_0, T] \subset \mathbb{R}$. The unknowns are the *(reduced) discrete scalar potentials* $\phi_\mathrm{u} : \mathcal{I} \to \mathbb{R}^{n_\phi}$, the *(reduced) discrete vector potentials* $\widehat{\mathrm{a}}_\mathrm{u} : \mathcal{I} \to \mathbb{R}^{n_\mathrm{a}}$, the auxiliary vector $\widehat{\pi}_\mathrm{u} : \mathcal{I} \to \mathbb{R}^{n_\pi}$ and the current $\mathrm{j_E} : \mathcal{I} \to \mathbb{R}^{n_\mathrm{E}}$ through the conductive contacts.

Remark 3.30. Let $(\phi_\mathrm{u}, \widehat{\mathrm{a}}_\mathrm{u}, \widehat{\pi}_\mathrm{u}) \in \mathbb{R}^{n_\phi} \times \mathbb{R}^{n_\mathrm{a}} \times \mathbb{R}^{n_\pi}$ be a solution of (3.60). Then all field quantities can be derived. We obtain

$$\widehat{\mathrm{b}}_\mathrm{u} = \mathrm{C}_\mathrm{u} \widehat{\mathrm{a}}_\mathrm{u},$$

$$\widehat{\mathrm{e}}_\mathrm{u} = -\mathrm{G}_\mathrm{u} \phi_\mathrm{u} + \Lambda_\mathrm{u} \mathrm{v_E} - \widehat{\pi}_\mathrm{u},$$

$$\widehat{\widehat{\mathrm{d}}}_\mathrm{u} = \mathrm{M}_\varepsilon^\mathrm{u} \widehat{\mathrm{e}}_\mathrm{u},$$

$$\widehat{\widehat{\mathrm{j}}}_{c,\mathrm{u}} = \mathrm{M}_\sigma^\mathrm{u} \widehat{\mathrm{e}}_\mathrm{u},$$

$$\widehat{\mathrm{h}}_\mathrm{u} = \mathrm{M}_\nu^\mathrm{u} (\widehat{\mathrm{b}}_\mathrm{u}) \widehat{\mathrm{b}}_\mathrm{u}$$

and

$$\mathrm{q}_\mathrm{u} = \widetilde{\mathrm{S}}_\mathrm{u} \widehat{\widehat{\mathrm{d}}}_\mathrm{u}.$$

3.2.6 Numerical Analysis of Maxwell's Grid Equations

In this subsection we investigate MGE (3.60) using the Coulomb and Lorenz gauge without the current equation since it is only an explicit function evaluation. We are mainly interested in the index of the resulting DAEs. We obtain similar results as [BCS12] but we do not use the differentiation index, the excitation of the fields is coming from boundary conditions instead of source term and we regard a different class of gauge conditions.

We collect some basic assumptions and properties for the discrete operators from the previous section.

Assumption 3.31. The reduced discrete conductivity matrix M_σ^u is a symmetric positive semi-definite diagonal matrix and M_ε^u is a symmetric positive definite diagonal matrix. Furthermore the reduced discrete material matrices M_ζ^u, M_ξ^u, $M_\nu^u(C_u\widehat{a}_u)$ and the reduced discrete differential reluctivity matrix $M_{\nu,d}^u(C_u\widehat{a}_u)$ are positive definite, see Remark 3.33.

Property 3.32. Let Assumption 3.28 be fulfilled. We have:

- $M_{\overline{\nu}}^u = M_\zeta^u G_u M_\xi^u \widetilde{S}_u M_\zeta^u$

- G_u has full column rank

- $C_u \Lambda_u$ has full column rank

- $G_u = -\widetilde{S}_u^T$, $\widetilde{G}_u = -S_u^T$ and $\widetilde{C}_u = C_u^T$

- $\widetilde{S}_u \widetilde{C}_u = 0$ and $C_u G_u = 0$

In the following we suppose that Assumption 3.31 and Property 3.32 are valid.

Maxwell's Grid Equations using Coulomb Gauge

First we focus a Coulomb gauge, that is, $\vartheta = 0$, and we obtain a DAE of the type

$$A(y,t)\frac{d}{dt}d(y,t) + b(y,t) = 0 \tag{3.61}$$

with

$$A = \begin{bmatrix} 0 & 0 \\ I & 0 \\ 0 & I \end{bmatrix}, \quad d(y,t) = \begin{pmatrix} M_\varepsilon^u G_u \phi_u + M_\varepsilon^u \widehat{\pi}_u \\ \widehat{a}_u \end{pmatrix}$$

and

$$b(y,t) = \begin{pmatrix} \widetilde{S}_u M_{\overline{\nu}}^u \widehat{a}_u \\ M_\sigma^u G_u \phi_u + K_\nu^u(\widehat{a}_u)\widehat{a}_u + M_\sigma^u \widehat{\pi}_u + \frac{d}{dt}M_\varepsilon^u G_b \phi_b + M_\sigma^u G_b \phi_b \\ -\widehat{\pi}_u \end{pmatrix},$$

where $K_\nu^u(\widehat{a}_u) = \widetilde{C}_u M_\nu^u(C_u\widehat{a}_u)C_u$. The DAE (3.61) has a properly stated leading term. With

$$D(y,t) = \begin{bmatrix} M_\varepsilon^u G_u & 0 & M_\varepsilon^u \\ 0 & I & 0 \end{bmatrix}.$$

it is easy to verify and we can choose $R = I$.

The next steps are: First we determine the higher index components. With that it is easy to show that the index is always greater than one. Finally we show that the index is always two. Following the index analysis we present an approach to compute suitable starting values for the numerical integration.

We determine the index of the DAE (3.61). We start with the first matrix of the matrix chain, see Definition 2.21, given by

$$G_0(y,t) = \begin{bmatrix} 0 & 0 & 0 \\ M^u_\varepsilon G_u & 0 & M^u_\varepsilon \\ 0 & I & 0 \end{bmatrix}.$$

Obviously the matrix $G_0(y,t)$ is always singular and thus the DAE (3.61) has not index-0, see Lemma 2.30. A projector onto $\ker G_0(y,t)$ is given by

$$Q_0 = \begin{bmatrix} I & 0 & 0 \\ 0 & 0 & 0 \\ -G_u & 0 & 0 \end{bmatrix}.$$

For the matrix chain we need the derivative of $b(y,t)$ with respect to the unknowns.

Remark 3.33 ([DMW08, Sch11]). The derivative of $K^u_\nu(\widehat{a}_u) = \tilde{C}_u M^u_\nu (C_u \widehat{a}_u) C_u$ with respect to $\widehat{a}_u$ is given by

$$K^u_{\nu,d}(\widehat{a}_u) = \tilde{C}_u M^u_{\nu,d}(C_u \widehat{a}_u) C_u$$

with:

$$\begin{aligned}
\frac{d}{d\widehat{a}_u}\left[\tilde{C}_u M^u_\nu(C_u \widehat{a}_u) C_u \widehat{a}_u\right] &= \frac{d}{d\widehat{a}_u}\left[\tilde{C}_u M^u_\nu(\widehat{b}_u)\widehat{b}_u\right] \\
&= \tilde{C}_u \frac{d}{d\widehat{b}_u}\left[M^u_\nu(\widehat{b}_u)\widehat{b}_u\right]\frac{d}{d\widehat{a}_u}\widehat{b}_u \\
&= \tilde{C}_u \frac{d}{d\widehat{b}_u}\left[M^u_\nu(\widehat{b}_u)\widehat{b}_u\right]\frac{d}{d\widehat{a}_u}[C_u \widehat{a}_u] \\
&= \tilde{C}_u \frac{d}{d\widehat{b}_u}\left[M^u_\nu(\widehat{b}_u)\widehat{b}_u\right] C_u \\
&= \tilde{C}_u M^u_{\nu,d}(C_u \widehat{a}_u) C_u
\end{aligned}$$

For Brauer's model the reduced discrete differential reluctivity matrix $M^u_{\nu,d}(C_u \widehat{a}_u)$ is positive define, see Corollary A.13. in [Sch11].

Then we get

$$B_0(y,t) = \begin{bmatrix} 0 & \tilde{S}_u M^u_\nu & 0 \\ M^u_\sigma G_u & K^u_{\nu,d}(\widehat{a}_u) & M^u_\sigma \\ 0 & 0 & -I \end{bmatrix}$$

and

$$B_0\left(y,t\right)Q_0 = \begin{bmatrix} 0 & 0 & 0 \\ 0 & 0 & 0 \\ G_u & 0 & 0 \end{bmatrix}.$$

The next step is the calculation of the intersection of $\mathcal{N}_0$ and $\mathcal{S}_0\left(y,t\right)$. That intersection is crucial for the index and for the consistent initialization as well. The intersection of $\mathcal{N}_0$ and $\mathcal{S}_0\left(y,t\right)$ can be described as follows.

Lemma 3.34. The Assumption 3.31 and Property 3.32 holds true. The index-1 set of the DAE (3.61) can be described by

$$\mathcal{N}_0 \cap \mathcal{S}_0\left(y,t\right) = \operatorname{im} Q_0$$

for all $\left(y,t\right) \in \mathcal{D} \times \mathcal{I}.$

Proof. For calculating the index-1 set we make use of Remark 2.27. For a suitable description we need a projector along $\operatorname{im} G_0\left(y,t\right)$. In order to determine such a projector we calculate a projector onto $\ker G_0\left(x,t\right)^\top$, see Remark A.8, with

$$G_0\left(y,t\right)^\top = \begin{bmatrix} 0 & -\widetilde{S}_u M_\varepsilon^u & 0 \\ 0 & 0 & I \\ 0 & M_\varepsilon^u & 0 \end{bmatrix}.$$

We can choose a projector onto $\ker G_0\left(y,t\right)^\top$ and along $\operatorname{im} G_0\left(y,t\right)$ by

$$W_0^\top = \begin{bmatrix} I & 0 & 0 \\ 0 & 0 & 0 \\ 0 & 0 & 0 \end{bmatrix} \text{ and } W_0 = \begin{bmatrix} I & 0 & 0 \\ 0 & 0 & 0 \\ 0 & 0 & 0 \end{bmatrix}.$$

We get

$$W_0 B_0\left(y,t\right)Q_0 = \begin{bmatrix} 0 & 0 & 0 \\ 0 & 0 & 0 \\ 0 & 0 & 0 \end{bmatrix}$$

and hence $\mathcal{N}_0 \cap \mathcal{S}_0\left(y,t\right) = \operatorname{im} Q_0.$ $\qquad\square$

It is obvious that the index-1 set $\mathcal{N}_0 \cap \mathcal{S}_0\left(y,t\right)$ is always not empty, that is, the DAE (3.61) has never index-1, see Definition 2.23. But the index does not exceed two as we will see in the next theorem.

Theorem 3.35 (index-2). Let Assumption 3.31 and Property 3.32 be fulfilled. The DAE (3.61) has index-2.

Proof. At first we need

$$G_1(y, t) = \begin{bmatrix} 0 & 0 & 0 \\ M_\varepsilon^u G_u & 0 & M_\varepsilon^u \\ G_u & I & 0 \end{bmatrix}$$

in order to proceed the matrix chain. For the characterization of the index-2 set we introduce a projector along $\operatorname{im} G_1(y, t)$, see Remark 2.28. On this we determine a projector onto $\ker G_1(x, t)^\top$, see Remark A.8. By investigating in

$$G_1(y, t)^\top = \begin{bmatrix} 0 & -\widetilde{S}_u M_\varepsilon^u & -\widetilde{S}_u \\ 0 & 0 & I \\ 0 & M_\varepsilon^u & 0 \end{bmatrix}.$$

We can choose a projector onto $\ker G_1(y, t)^\top$ and along $\operatorname{im} G_1(y, t)$ by $W_1^\top = W_0^\top$ and $W_1 = W_0$. Next we take into account

$$P_0 = \begin{bmatrix} 0 & 0 & 0 \\ 0 & I & 0 \\ G_u & 0 & I \end{bmatrix}, \quad B_0(y, t) P_0 = \begin{bmatrix} 0 & \widetilde{S}_u M_\nu^u & 0 \\ M_\sigma^u G_u & K_{\nu,d}^u(\widehat{a}_u) & M_\sigma^u \\ -G_u & 0 & -I \end{bmatrix},$$

where P_0 is the complementary projector to Q_0, and

$$W_1 B_0(y, t) P_0 = \begin{bmatrix} 0 & \widetilde{S}_u M_\nu^u & 0 \\ 0 & 0 & 0 \\ 0 & 0 & 0 \end{bmatrix}.$$

Let be $z \in \ker G_1(y, t) \cap \ker W_1 B_0(y, t) P_0$. That is true if and only if the conditions

$$z_{\widehat{\pi}_u} = -G_u z_{\phi_u} \tag{3.62}$$

$$z_{\widehat{a}_u} = -G_u z_{\phi_u} \tag{3.63}$$

$$\widetilde{S}_u M_\nu^u z_{\widehat{a}_u} = 0 \tag{3.64}$$

are fulfilled. Left-multiplying (3.63) by $\widetilde{S}_u M_\nu^u$ and using (3.64) yields

$$\widetilde{S}_u M_\nu^u G_u z_{\phi_u} = 0$$

and hence $z_{\phi_u} = 0$ due to the choice of $M_\nu^u = M_\zeta^u G_u M_\xi^u \widetilde{S}_u M_\zeta^u$. From (3.62) and (3.63) we get $\left(z_{\widehat{a}_u}, z_{\widehat{\pi}_u}\right) = 0$ and conclude $z = 0$, see Definition 2.23. $\qquad\square$

In order to start the integration of the DAE (3.61) we need a consistent initialization. For the index-2 case we apply Theorem 2.58.

Assumption 3.36. For the DAE (3.61) exists the continuous partial derivatives $\frac{\partial}{\partial t} d(y, t)$ and $\frac{\partial}{\partial t} W_1 b(y, t)$ for all $(y, t) \in \mathcal{D} \times \mathcal{I}$.

These assumptions are not a restriction since, if a solution exists, then $\frac{\partial}{\partial t}d(y,t)$ exists and is continuous. Moreover $W_1 b(y,t)$ describes exactly the hidden constraints and hence $\frac{\partial}{\partial t} W_1 b(y,t)$ needs to exists and to be continuous to have a solution to the problem. In addition the DAE (3.61) has a constant matrix A and there are the constant projectors Q_0 and W_1. We need to show that the index-2 variables enter linearly only.

Lemma 3.37. Let Assumption 3.31 and Property 3.32 be fulfilled. The index-2 variables enter the DAE (3.61) linearly only.

Proof. From Lemma 3.34 we easily obtain a constant projector T onto $\mathcal{N}_0 \cap \mathcal{S}_0(y,t)$ given by Q_0 and $U = P_0$. The unknowns are divided into

$$y = Ty + Uy = \begin{pmatrix} \phi_u \\ 0 \\ -G_u\phi_u \end{pmatrix} + \begin{pmatrix} 0 \\ \widehat{a}_u \\ G_u\phi_u + \widehat{\pi}_u \end{pmatrix}.$$

Now we can write $b(y,t) = b(Uy,t) + BTy$ with

$$B = \begin{bmatrix} 0 & 0 & 0 \\ 0 & 0 & 0 \\ 0 & 0 & -I \end{bmatrix}.$$

The relation $d(y,t) = d(Uy,t)$ is obvious by Lemma 2.54. □

The DAE (3.61) fulfills all requirements to apply Theorem 2.58 in case of index-2. But we still need an operating point when we want to integrate it numerically. Since Theorem 2.58 is applicable to the DAE an operating point is sufficient to start the numerical integration, see Lemma 2.61.

Lemma 3.38. Let the DAE (3.61) be given and $t_0 \in \mathcal{I}$. An operating point (z^0, y^0, t_0) with $z^0 = \left(z^0_{\widehat{a}_u}, z^0_{\widehat{\pi}_u}\right)$ and $y^0 = (\phi^0_u, \widehat{a}^0_u, \widehat{\pi}^0_u)$ can be calculated as follows:

- Choose $\phi^0_u \in \mathbb{R}^{n_\phi}$ and $\widehat{\pi}^0_u \in \mathbb{R}^{n_\pi}$ arbitrarily, and $\widehat{a}^0_u \in \ker \widetilde{S}_u M^u_{\widetilde{\nu}}$.

- Compute the missing parts by:

$$M^u_\varepsilon z^0_{\widehat{a}_u} = M^u_\sigma G_u \phi^0_u + K^u_\nu(\widehat{a}^0_u)\widehat{a}^0_u + M^u_\sigma \widehat{\pi}^0_u + \frac{d}{dt}M^u_\varepsilon G_b \phi_b(t_0) + M^u_\sigma G_b \phi_b(t_0)$$
$$z^0_{\widehat{\pi}_u} = \widehat{\pi}^0_u$$

□

Remark 3.39. Due to the structure of the DAE (3.61) we obtain a locally unique solution through every consistent initial value and perturbation index-2, see Theorem 2.46, and 2.50.

Maxwell's Grid Equations using Lorenz Gauge

Next we take Lorenz gauge into account, that is $\vartheta = 1$. Let Assumption 3.31 and Property 3.32 be fulfilled. Then we obtain an ODE of the form

$$A \frac{d}{dt} y + b(y, t) = 0 \qquad (3.65)$$

with

$$A = \begin{bmatrix} \widetilde{S}_u M_\varepsilon^u G_u & 0 & 0 \\ M_\varepsilon^u G_u & 0 & M_\varepsilon^u \\ 0 & I & 0 \end{bmatrix}$$

and

$$b(y, t) = \begin{pmatrix} \widetilde{S}_u M_{\bar{\nu}}^u \widehat{a}_u \\ M_\sigma^u G_u \phi_u + K_\nu^u(\widehat{a}_u) \widehat{a}_u + M_\sigma^u \widehat{\pi}_u + \frac{d}{dt} M_\varepsilon^u G_b \phi_b + M_\sigma^u G_b \phi_b \\ -\widehat{\pi}_u \end{pmatrix}.$$

Lemma 3.40. Let Assumption 3.31 and Property 3.32 be fulfilled. Then, MGE (3.60) using Lorenz gauge is an ODE of the form (3.65).

Proof. From Assumption 3.31 and Property 3.32 we deduce that $\widetilde{S}_u M_\varepsilon^u G_u$ and M_ε^u are nonsingular. Thus,

$$A = \begin{bmatrix} \widetilde{S}_u M_\varepsilon^u G_u & 0 & 0 \\ M_\varepsilon^u G_u & 0 & M_\varepsilon^u \\ 0 & I & 0 \end{bmatrix}$$

is nonsingular. $\qquad \square$

Hence for Lorenz gauge we have no restriction for initial values.

Remark 3.41. The chosen gauge condition for MGE (3.60) has a huge impact on the structure of the resulting system. In case of the Coulomb gauge we obtain an index-2 DAE and for Lorenz gauge we attain an ODE. That is, from the numerical point of view Lorenz gauge is to be prefer. Next we consider the Jacobians results from integrating the DAE in time by BDF methods with step size $h > 0$. For the DAE (2.30) the Jacobian reads

$$J(y, t) = \frac{\alpha_0}{h} A(t) D(y, t) + B_0(y, t).$$

Depending on the choice of M_ζ^u and M_ξ^u, the MGE (3.60) using Lorenz gauge may leads to more dense Jacobians than using Coulomb gauge. In addition the structure of the Jacobians depending on the gauge. For Lorenz gauge the Jacobian reads

$$J_L(y, t) = \begin{bmatrix} \frac{\alpha_0}{h} \widetilde{S}_u M_\varepsilon^u G_u & \widetilde{S}_u M_{\bar{\nu}}^u & 0 \\ \left(M_\sigma^u + \frac{\alpha_0}{h} M_\varepsilon^u \right) G_u & K_{\nu,d}^u(\widehat{a}_u) & \left(M_\sigma^u + \frac{\alpha_0}{h} M_\varepsilon^u \right) \\ 0 & \frac{\alpha_0}{h} I & -I \end{bmatrix}$$

with a nonzero diagonal and for Coulomb gauge

$$
J_C\left(y,t\right) = \begin{bmatrix} 0 & \widetilde{S}_u M_{\bar{\nu}}^u & 0 \\ \left(M_\sigma^u + \frac{\alpha_0}{h} M_\varepsilon^u\right) G_u & K_{\nu,d}^u(\widehat{a}_u) & \left(M_\sigma^u + \frac{\alpha_0}{h} M_\varepsilon^u\right) \\ 0 & \frac{\alpha_0}{h} I & -I \end{bmatrix}.
$$

Hence the Lorenz gauge system could be suitable for iterative solvers, particularly with regard to the possible large number of unknowns. Note that for sufficient small $h > 0$ the Jacobian $J_L\left(y,t\right)$ and $J_C\left(y,t\right)$ are nonsingular due to Lemma 2.6.

Remark 3.42. Coulomb gauge could be suitable for iterative solvers, too. For this we need to reformulate the DAE (3.61). We add the Coulomb gauge (3.56), $\vartheta = 0$, by a grad-div formulation directly to the reduced discrete Maxwell-Ampère's law (3.51) and we add the reduced discrete continuity equation (3.54) to the system equations. This leads to the DAE

$$
\widetilde{S}_u M_\varepsilon^u G_u \frac{d}{dt} \phi_u + \widetilde{S}_u M_\varepsilon^u \frac{d}{dt} \widehat{\pi}_u + \widetilde{S}_u M_\sigma^u G_u \phi_u + \widetilde{S}_u M_\sigma^u \widehat{\pi}_u
$$
$$
- \widetilde{S}_u M_\varepsilon^u \Lambda_u \frac{d}{dt} v_E - \widetilde{S}_u M_\sigma^u \Lambda_u v_E = 0
$$
$$
M_\varepsilon^u G_u \frac{d}{dt} \phi_u + M_\varepsilon^u \frac{d}{dt} \widehat{\pi}_u + M_\sigma^u G_u \phi_u + \widetilde{C}_u M_\nu^u (C_u \widehat{a}_u) C_u \widehat{a}_u + M_\zeta^u G_u M_\xi^u \widetilde{S}_u M_\zeta^u \widehat{a}_u + M_\sigma^u \widehat{\pi}_u
$$
$$
- M_\varepsilon^u \Lambda_u \frac{d}{dt} v_E - M_\sigma^u \Lambda_u v_E = 0
$$
$$
\frac{d}{dt} \widehat{a}_u - \widehat{\pi}_u = 0
$$

where the Coulomb gauge is implicitly fulfilled. The BDF Jacobian for this DAE is given by

$$
J_C\left(y,t\right) = \begin{bmatrix} \widetilde{S}_u \left(M_\sigma^u + \frac{\alpha_0}{h} M_\varepsilon^u\right) G_u & 0 & \widetilde{S}_u \left(M_\sigma^u + \frac{\alpha_0}{h} M_\varepsilon^u\right) \\ \left(M_\sigma^u + \frac{\alpha_0}{h} M_\varepsilon^u\right) G_u & K_{\nu,d}^u(\widehat{a}_u) + M_\zeta^u G_u M_\xi^u \widetilde{S}_u M_\zeta^u & \left(M_\sigma^u + \frac{\alpha_0}{h} M_\varepsilon^u\right) \\ 0 & \frac{\alpha_0}{h} I & -I \end{bmatrix}
$$

with a nonzero diagonal.

Remark 3.43. It seems that MGE (3.60) using Coulomb gauge has some disadvantages compared to Lorenz gauge. However, a reformulation of MGE (3.60) using Coulomb gauge with $M_\zeta^u = M_\varepsilon^u$ proposed by [Jan12b] lead to an ODE if we disregard the current equation and taking into account that $\widetilde{S}_u M_\zeta^u G_u$ and M_ε^u are nonsingular. The idea is to exploit the kernel of the Coulomb gauge. Let $\{b_1, \ldots, b_k\}$ be an orthonormal basis with respect to the standard scalar product on $\mathbb{R}^k$ of $\ker \widetilde{S}_u M_\varepsilon^u$. Moreover let $\{b_{k+1}, \ldots, b_{n_a}\}$ be an orthonormal extension of $\{b_1, \ldots, b_k\}$ to an orthonormal basis with respect to the standard scalar product on $\mathbb{R}^{n_a}$. We define $B_P = \begin{bmatrix} b_{k+1} & \cdots & b_{n_a} \end{bmatrix} \in \mathbb{R}^{n_a \times n_a - k}$. Then $P = B_P B_P^\top$ is a projector along $\ker \widetilde{S}_u M_\varepsilon^u$ and we obtain

$$
0 = \widetilde{S}_u M_\varepsilon^u \widehat{a}_u = \widetilde{S}_u M_\varepsilon^u P \widehat{a}_u = \widetilde{S}_u M_\varepsilon^u B_P B_P^\top \widehat{a}_u.
$$

Due to the choice of B_P we obtain that $\widetilde{S}_u M_\varepsilon^u B_P$ is nonsingular and hence

$$B_P^\top \widehat{a}_u = 0.$$

With $Q = B_Q B_Q^\top$, $B_Q = \begin{bmatrix} b_1 & \cdots & b_k \end{bmatrix} \in \mathbb{R}^{n \times k}$, and $\widehat{a}_q = B_Q^\top \widehat{a}_u$, $\widehat{\pi}_q = B_Q^\top \widehat{\pi}_u$ we obtain

$$M_\varepsilon^u G_u \frac{\mathrm{d}}{\mathrm{d}t} \phi_u + M_\varepsilon^u B_Q \frac{\mathrm{d}}{\mathrm{d}t} \widehat{\pi}_q + M_\sigma^u G_u \phi_u + K_\nu^u (B_Q \widehat{a}_q) B_Q \widehat{a}_q + M_\sigma^u B_Q \widehat{\pi}_q$$

$$-M_\varepsilon^u \Lambda_u \frac{\mathrm{d}}{\mathrm{d}t} v_E - M_\sigma^u \Lambda_u v_E = 0 \qquad (3.66)$$

$$\frac{\mathrm{d}}{\mathrm{d}t} \widehat{a}_q - \widehat{\pi}_q = 0$$

from MGE (3.60). We split the first equation of (3.66) using $B_Q^\top$ and $B_P^\top$. We achieve

$$B_Q^\top M_\varepsilon^u B_Q \frac{\mathrm{d}}{\mathrm{d}t} \widehat{\pi}_q + B_Q^\top M_\sigma^u G_u \phi_u + B_Q^\top K_\nu^u (B_Q \widehat{a}_q) B_Q \widehat{a}_q + B_Q^\top M_\sigma^u B_Q \widehat{\pi}_q$$

$$-B_Q^\top M_\varepsilon^u \Lambda_u \frac{\mathrm{d}}{\mathrm{d}t} v_E - B_Q^\top M_\sigma^u \Lambda_u v_E = 0$$

$$B_P^\top M_\varepsilon^u G_u \frac{\mathrm{d}}{\mathrm{d}t} \phi_u + B_P^\top M_\varepsilon^u B_Q \frac{\mathrm{d}}{\mathrm{d}t} \widehat{\pi}_q + B_P^\top M_\sigma^u G_u \phi_u + B_P^\top K_\nu^u (B_Q \widehat{a}_q) B_Q \widehat{a}_q + B_P^\top M_\sigma^u B_Q \widehat{\pi}_q$$

$$-B_P^\top M_\varepsilon^u \Lambda_u \frac{\mathrm{d}}{\mathrm{d}t} v_E - B_P^\top M_\sigma^u \Lambda_u v_E = 0$$

$$\frac{\mathrm{d}}{\mathrm{d}t} \widehat{a}_q - \widehat{\pi}_q = 0$$

using Property 3.32. Since $B_Q^\top M_\varepsilon^u B_Q$ and $B_P^\top M_\varepsilon^u G_u$ are nonsingular we obtain an ODE for $(\phi_u, \widehat{a}_q, \widehat{\pi}_q)$. In fact, that is some kind of index reduction using knowledge of the solution of $\widehat{a}_u$ given by the Coulomb gauge.

3.3 Summary

This chapter has briefly introduced Maxwell's equations and the finite integration technique for the resulting spatial discretization. We discussed a potential formulation of Maxwell's equations and presented a general class of gauge conditions. Next we motivated Dirichlet boundary conditions for the potentials.

General properties of the discrete operators in terms of the finite integration technique were discussed. The Maxwell's grid equations (3.60) were formulated in terms of potentials with incorporated boundary conditions using a new class of discrete gauge conditions (3.55) in terms of the finite integration technique. We defined a suitable boundary excitation and formulated current equations (3.58) for the currents through the electromagnetic devices to be easily accessible for circuit simulation. The chosen approach differs substantially from [DHW04, Ben06, BBS11, Sch11], where the excitation is constructed using several conductor models and applied as a source term.

The structural properties of Maxwell's grid equations (3.60) formulated as a differential-algebraic equation with a properly stated leading term were discussed and analyzed by the index concept to obtain new index results. The first new result was that the index depends on the chosen gauge condition. The Coulomb gauge leads to the locally unique solvable index-2 and perturbation index-2 differential-algebraic equation (3.61) formulated with a properly stated leading term (Theorem 3.35 and Remark 3.39) with linear index-2 variables (Lemma 3.37) and we provided a way to calculate an operating point (Lemma 3.38) to determine a consistent initialization. Maxwell's grid equations turned out to be an ordinary differential equation (3.65) using Lorenz gauge (Lemma 3.40). These results were obtained without taking the currents through the device into account.

We analyzed the structural differences of Maxwell's grid equations using Coulomb and Lorenz gauge (Remark 3.41 and 3.42). Finally, we reformulated Maxwell's grid equations using a particular Coulomb gauge as an ordinary differential equation (3.66) by an orthonormal basis decomposition (Remark 3.43) without taking the currents through the device into account. From the results of both ordinary differential equations it can be concluded that the modeling of the Maxwell's grid equations has an impact on the perturbation sensitivity and thus careful modeling is desirable.

4 Electric Network

Today electric networks and circuits are indispensable. They can be found in almost every electronic device from radios to central processing units of our personal computer to smartphones. An electric network is the interconnection of elements such as condensers, resistors, coils and batteries modeled by capacitors, resistors, inductors, current sources and voltage sources or more complex elements such as diodes and metal-oxide-semiconductor field-effect transistor.

To reduce cost and development cycles of new electric products numerical simulations are used to predict the circuit's behavior in terms of physical quantities such as voltages and currents. A suitable model for numerical simulation of electric networks has to meet two contradicting requirements. On the one hand the physical behavior of an electric network should be as correct as possible. On the other hand the model has to be simple enough to keep the simulation time reasonably small. With regard to the simulation time usually the first step is to restrict the circuit elements to the *basic elements* capacitors, resistors, inductors, current sources and voltage sources while other elements are replaced by *equivalent circuits*, that is, basic elements only.

A well-established modeling approach to meet the requirements is the modified nodal analysis providing a system with a relatively small dimension that is able to automatically setup the network equations, see [CL75, CDK87, DK84]. This model analysis is successfully applied in established programs such as SPICE (Electronics Research Laboratory of the University of California, Berkeley) and TITAN (Infineon Technologies AG).

For today's challenges the circuit industry is continuously developing new circuits and circuit elements. In 2008 HP Labs announced the physical realization of a new circuit element, namely, the memristor, whose existence was postulated in 1971 by Leon Chua, see [Chu71, SSSW08]. This has motivated further research on memristors since many potential applications are reported such as storing huge amount of data or replacing transistors. The use of memristors in circuit simulation requires some effort and the memristor needs to be embedded in actual circuit models. The nodal analysis method has already been extended by memristor models. The index of the resulting differential-algebraic equation is investigated in [Ria10]. In this thesis we extend the modified nodal analysis by memristor models and investigate the structural properties of resulting differential-algebraic equation formulated with a properly stated leading term.

This chapter is organized as follows. First, we introduce the characteristic equations and topology for the basic circuit elements known from literature, [CDK87, DK84], and, in addition, for the memristor, [Chu71]. Next, we familiarize with the modified nodal analysis. Finally we extend the modified nodal analysis by memristor models and the resulting differential-algebraic equation with a properly stated leading term is analyzed in terms of the index. We extend the well-known topological index conditions of [Tis99, ET00] for the modified nodal analysis to circuits including memristors. In addition, we present an approach to calculate a consistent initialization for the modified nodal analysis including memristor models.

4.1 Network Modeling

ME are also applicable to circuits. However, the complexity of integrated circuits makes simplifications unavoidable. Therefore an independent theory was deduced from ME, tailored for circuit simulations, see [CL75].
The spatial dimensions of the elements are disregarded in this investigations. Two preconditions must be met: The electrical connections between the circuit elements have to be ideally conducting and the nodes have to be ideal and concentrated. The physical behavior of the circuit elements is modeled by characteristic equations.

In the modified nodal analysis (MNA - Modified Nodal Analysis) the circuit is modeled by a network graph and the topology can be described by Kirchhoff's laws, see [DK84, Ria08]. We restrict ourselves to elements with two contacts and terminals, respectively, that is, every circuit element is represented by an edge with a different front and back node.

4.1.1 Basic Electric Elements

The physical behavior of each network element is modeled by the relation between its edge currents and its edge voltages.
We specify the characteristic equations for the basic elements, that is for capacitors, resistors, inductors, voltage and current sources, in terms of currents and voltages through the elements, see [CL75, CDK87]. A part from the sources characteristic equations are deducible from ME by neglecting certain effects. Capacitors store energy in their electric

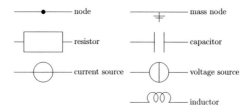

Figure 4.1: Symbols of circuit elements.

field. The electric charges of the capacitor are modeled by a function $q_C : \mathbb{R}^{n_C} \times \mathcal{I} \to \mathbb{R}^{n_C}$ and the characteristic equations are given by

$$j_C = \frac{d}{dt} q_C \left(v_C, t \right),$$

where $j_C, v_C : \mathcal{I} \to \mathbb{R}^{n_C}$ are the capacitors currents, voltages and $n_C \in \mathbb{N}$ is the number of capacitors and $\mathcal{I} \subset \mathbb{R}$.

Resistors limit the flow of electrical current by generating voltage drops and may be described by a function $g_R : \mathbb{R}^{n_R} \times \mathcal{I} \to \mathbb{R}^{n_R}$ given by

$$j_R = g_R \left(v_R, t \right),$$

where $j_R, v_R : \mathcal{I} \to \mathbb{R}^{n_R}$ are the resistor currents and voltages and $n_R \in \mathbb{N}$ is the number of resistors.

Inductors store energy in their magnetic field. The magnetic flux of the inductors is modeled by the function $\phi_L : \mathbb{R}^{n_L} \times \mathcal{I} \to \mathbb{R}^{n_L}$ and the characteristic equation is given by

$$v_L = \frac{d}{dt} \phi_L \left(j_L, t \right)$$

with $j_L, v_L : \mathcal{I} \to \mathbb{R}^{n_L}$ being the inductor currents and voltages and $n_L \in \mathbb{N}$ the number of inductors.

We confine our investigation to independent sources. Voltage and current sources are distinguished by the fact that the voltage and the current are given by

$$v_V = v_s \left(t \right) \text{ and } j_I = i_s \left(t \right)$$

with $v_V : \mathcal{I} \to \mathbb{R}^{n_V}$ and $j_I : \mathcal{I} \to \mathbb{R}^{n_I}$, where $n_V, n_I \in \mathbb{N}$ is the number of voltage and current sources.

4.1.2 Memristors

If it's pinched, it's a memristor.

LEON CHUA ABOUT THE CHARACTERIZATION OF
A RESISTANCE MEMORY DEVICE, [CHU11].

In 1971 Leon Chua introduced a new circuit element named *memristor* [Chu71]. He motivated the plausibility that such a device might someday be discovered by ME, see [Chu71, AASE+10]. This element provides a nonlinear relationship between the charge and the flux and hence it completes the conceptual symmetry with the resistor, whose characteristic relate current and voltage, the inductor, involving current and flux, and the capacitor, which relates voltage and charge. In 2008, a physical model of a two-contact device behaving like a memristor was announced in [SSSW08]. This has motivated a lot of research on this topic, and the memristor and related devices are likely to have a great impact on electronics in the near future at the nanometer scale, see references in

Figure 4.2: Symbols of memristor elements.

[Ria10, Ria11]. For this reason the memristor needs to be embedded in actual circuit models.

Memristors are governed by charge-flux relations $g_M : \mathbb{R}^{n_M} \times \mathbb{R}^{n_M} \times \mathcal{I} \to \mathbb{R}^{n_M}$ of the type

$$g_M(\phi, q_M, t) = 0$$

with $\phi, q_M : \mathcal{I} \to \mathbb{R}^{n_M}$, $n_M \in \mathbb{N}$ is the number of memristors and $\mathcal{I} \subset \mathbb{R}$. In the following we assume that the devices have two contacts and are either charge-controlled, that is, the fluxes can be expressed by

$$\phi = \phi_M(q_M, t),$$

where $\phi_M : \mathbb{R}^{n_M} \times \mathcal{I} \to \mathbb{R}^{n_M}$. We assume that the partial derivatives

$$M(q, t) = \frac{\partial}{\partial q} \phi_M(q, t)$$

exist and is continuous. We call $M : \mathbb{R}^{n_M} \times \mathcal{I} \to \mathbb{R}^{n_M \times n_M}$ the *memristance*. Together with the basic relations

$$\frac{\mathrm{d}}{\mathrm{dt}} \phi_M(q_M, t) = v_M \text{ and } \frac{\mathrm{d}}{\mathrm{dt}} q_M = j_M, \tag{4.1}$$

where $j_M, v_M : \mathcal{I} \to \mathbb{R}^{n_M}$ are the memristors currents and voltages, we can conclude

$$v_M = M(q_M, t) j_M$$

and it becomes clear why that devices are called memristors. In case of a constant memristance the memristors do not differ from resistors. For a non-constant memristance the memristance depends on

$$q_M(t) = \int_{-\infty}^{t} j_M(\tau) \, \mathrm{d}\tau$$

and hence the memristors have an memory effect.

In [Ria10, RT11, Ria11] an extension of the nodal analysis and in [BT10, FY10] an extension of the MNA are presented including memristor models. There is a number of SPICE implementations of the memristor, see [BBB09a, KKS10, BBBK10, AASE$^+$10] and references therein. Most SPICE models of the memristor are developed on the basis of the HP memristor or using subcircuits to model the memristor's behavior. In [SSSW08, KKS10, BBBK10, Chu11] memristances are given.
A lot of recent research is focused on devices closely related to the memristor, such as the memcapacitors and meminductors recently introduced in [CPD09, BBB09b]. These and other related circuit elements are beyond the scope of the thesis.

4.1.3 Network Topology and Kirchhoff Laws

We model a circuit by a directed graph $\mathcal{G} := (\mathcal{N}, \mathcal{E})$ with arbitrarily orientation, see [DK84, Ria08]. Then the network topology for elements with two contacts is retained by the (reduced) incidence matrix $A \in \{-1, 0, 1\}^{n_N \times n_b}$, see Appendix B. The matrix A describes in an elegant way the relation between all $n_N + 1 = |\mathcal{N}|$ nodes and all $n_e = |\mathcal{E}|$ edges of the circuit. The (reduced) incidence is defined by:

$$(A)_{ij} = \begin{cases} 1 & \text{if the edge } j \text{ leaves node } i, \\ -1 & \text{if the edge } j \text{ enters node } i, \\ 0 & \text{else.} \end{cases}$$

The reference node is called *mass node* and is an arbitrarily node of $\mathcal{G}$.

A milestone for circuit modeling are the Kirchhoff's laws, which deal with the conservation of charge and energy in electrical circuits and were first described in 1845 by Gustav Kirchhoff. Both laws can be directly derived from ME, but Kirchhoff preceded Maxwell and instead generalized the work by Georg Ohm.

The Kirchhoff's laws take into account the circuit's topology:

(i) *Kirchhoff's voltage law* (KVL - <u>K</u>irchhoff's <u>V</u>oltage <u>L</u>aw): At every instant of time the algebraic sum of voltages along each loop of the network is equal to zero.

(ii) *Kirchhoff's current law* (KCL - <u>K</u>irchhoff's <u>C</u>urrent <u>L</u>aw): At every instant of time the algebraic sum of currents entering one node of the network is equal to zero.

KVL and KCL can be deduced from ME. We start from the following Assumptions: First, *cross talk*, that is, undesired capacitive, inductive, or conductive coupling from one circuit element to another, can be neglected. Second, there is no time evolution of the EM fields. Last, the electrical connections between the circuit elements to be ideally conducting and the nodes to be ideal and concentrated. If these assumptions are met ME imply the Kirchhoff's laws. KCL can be derived by the continuity equation (3.5) and KVL by Maxwell-Faraday law (3.3), respectively. In the static case ME leads to:

$$\nabla \cdot \vec{J}_c = 0 \tag{4.2}$$

$$\nabla \times \vec{E} = 0 \tag{4.3}$$

Applying Gauss' law to (4.2) we achieve

$$\int_F \vec{J}_c \cdot \vec{n}_\perp dF = \int_V \nabla \cdot \vec{J}_c dV = 0,$$

where V donates the volume and $F = \partial V$ the surface of an idealized electrical node. The current is defined by

$$i = \int_F \vec{J}_c dF.$$

Considering one node with edge currents $i_1, \ldots, i_m$ with $F = \sum_{i=1}^{m} F_i$, see Figure 4.3(a), entering this node we may describe KCL as

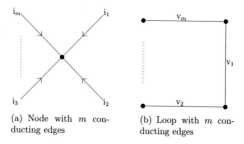

(a) Node with m conducting edges

(b) Loop with m conducting edges

Figure 4.3: KCL and KVL.

$$\sum_{k=1}^{m} i_k = \sum_{k=1}^{m} \int_{F_k} \vec{J}_c \cdot \vec{n}_\perp dF = \int_F \vec{J}_c \cdot \vec{n}_\perp dF = 0$$

that is, the sum of all edge currents entering a node equals zero. Applying Stokes' law on (4.3) we achieve

$$\int_F \vec{E} \cdot \vec{n}_\parallel dF = \int_E \nabla \times \vec{E} \cdot \vec{n}_\perp dE = 0$$

with $E = \partial F$ and F being a loop of idealized electrical wires. The voltage is defined by

$$v = \int_E \vec{E} \cdot \vec{n}_\parallel dE$$

If we consider a loop with the edge voltages $v_1, \ldots, v_m$ with $E = \sum_{i=1}^{m} E_i$, see Figure 4.3(b), then we can formulate KVL as

$$\sum_{k=1}^{m} v_k = \sum_{k=1}^{m} \int_{E_k} \vec{E} \cdot \vec{n}_\parallel dE = \int_E \vec{E} \cdot \vec{n}_\parallel dE = 0$$

that is, the sum of all edge voltages in a loop equals zero.

Let a connected electric network be given and $j, v \in \mathbb{R}^{n_b}$ be the vectors of all edge currents and voltages. Then KCL and KVL imply

$$Aj = 0 \tag{4.4}$$

and

$$v = A^\mathsf{T} e, \tag{4.5}$$

where $e \in \mathbb{R}^{n_N}$ are called node potentials, see [DK84], where the node potentials are defined as voltage drop with respect to the mass node. The node potentials leads to a smaller system size compared to a system with the edge voltages as variables. This is due to the fact that the network graph usually contains considerably more edges than nodes.

4.2 Modified Nodal Analysis for Circuits including Memristors

In this section we extend the charge oriented MNA, [FG99, ET00, Gün01], for circuits including memristors. We arrive at the system as [FY10], but in contrast to [FY10] we provide a detailed analysis of the resulting DAE.

The four essential steps in setting up the equations of the MNA equations are:

(i) Apply KCL to every node, except for the mass node, that is, start from (4.4).

(ii) Replace the characteristic equations for currents of resistors, capacitors and current sources in KCL.

(iii) Add the characteristic equations for inductors.

(iv) Add the characteristic equations for voltage sources and apply KVL (4.5) to obtain a formulation in node potentials instead of branch voltages.

The first step to gain structure information is sorting the network edges in such a way that the incidence matrix A forms a block matrix with blocks describing the different types of network elements, that is,

$$A = \begin{bmatrix} A_C & A_R & A_L & A_V & A_I \end{bmatrix},$$

where the index stands for capacitive, resistive, inductive, voltage source and current source edges, respectively, see [Tis99, ET00].

We are back to the MNA equations, which results in a DAE system of the form

$$A_C \frac{d}{dt} q_C \left(A_C^\top e, t \right) + A_R g_R \left(A_R^\top e, t \right) + A_L j_L + A_V j_V + A_I i_s \left(t \right) = 0$$

$$\frac{d}{dt} \phi_L \left(j_L, t \right) - A_L^\top e = 0 \qquad (4.6)$$

$$A_V^\top e - v_s \left(t \right) = 0$$

in time $t \in \mathcal{I}$, $\mathcal{I} = [t_0,\ T] \subset \mathbb{R}$. Denoting the number of nodes - except for the mass node - by n_N, the number of inductive edges by n_L and the number of voltage source edges by n_V. The dimension of the system is $n_N + n_L + n_V$. The given vector functions

$q_C(v,t)$, $g_R(v,t)$, $\phi_L(j,t)$, $v_s(t)$ and $i_s(t)$ describe the characteristic equations for the circuit elements.

The unknowns are the node potentials $e : \mathcal{I} \rightarrow \mathbb{R}^{n_N}$, except of the mass node, as well as the currents $j_L : \mathcal{I} \rightarrow \mathbb{R}^{n_L}$ through inductors and the currents $j_V : \mathcal{I} \rightarrow \mathbb{R}^{n_V}$ through voltage sources. The potential at the mass node is assigned to zero.

The first equation of (4.6) states KCL and the second one states the characteristic equations for inductances. The last equation combines the characteristic equations and KVL for the voltage sources. Details can be found in [ET00, Tis04].

Remark 4.1. In stating the model as we do we implicitly assume independent voltage and current sources only. For results with a broad class of controlled sources we refer to [ET00].

For the MNA (4.6) there are well-known index results depending on the circuit topology only, [Tis99, ET00]. For this we need the following assumptions and definitions.

Assumption 4.2 (no short circuit). The matrices A_V and $\begin{bmatrix} A_C & A_R & A_L & A_V \end{bmatrix}^T$ have full column rank, that is, it exists neither a loop containing only voltage sources (V-loop) nor a cutset containing only current sources (I-cutset), see Remark B.14.

These assumptions are necessary for a consistent model description and very natural since a violation would in reality result to a short circuit. From the mathematical point of view, the circuit equations would have either no solution or infinite many solutions due to KCL and KVL.

Example 4.3. The linear circuit in Figure 4.4(a) has a V-loop and the MNA (4.6) lead to

$$j_V^1 + j_V^2 + Ge = 0$$
$$e = v_s^1(t)$$
$$e = v_s^2(t)$$

with infinitely many solutions if and only if $v_s^1(t) = v_s^2(t)$, otherwise no solutions exist. The linear circuit in Figure 4.4(b) has an I-cutset and the MNA equations (4.6) lead to

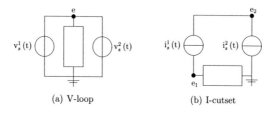

(a) V-loop   (b) I-cutset

Figure 4.4: Example of a V-loop and an I-cutset.

$$i_s^1(t) - i_s^2(t) = 0$$
$$i_s^1(t) + Ge_1 = 0$$

with infinitely many solutions if and only if $i_s^1(t) = i_s^2(t)$ since e_2 can be chosen freely, otherwise no solutions exists.

Assumption 4.4 (Passivity, [Bar04]). The functions $q_C(u,t)$, $\phi_L(j,t)$ and $g_R(u,t)$ are continuously differentiable with

$$C(u,t) = \frac{\partial}{\partial u}q_C(u,t), \; L(j,t) = \frac{\partial}{\partial j}\phi_L(j,t), \; G(u,t) = \frac{\partial}{\partial u}g_R(u,t),$$

being positive definite. Physically, we say that the elements are locally passive, that is, they do not produce any energy.

For the later analysis special loops and cutsets play a key role, [ET00].

Definition 4.5 (LI-cutset). A cutset is called LI-cutset if and only if the cutset contains only inductors and current sources.

Definition 4.6 (CV-loop). A loop is called CV-loop if and only if the loop contains only capacitors and voltage sources.

Theorem 4.7. Let Assumption 4.2 and Assumption 4.4 be fulfilled. The MNA (4.6) represent a DAE (2.18) with a properly stated leading term. The DAE has

- index-0 if and only if there are no voltage sources in the circuit and the circuit has a tree containing capacitors only,

- index-1 if and only if there is at least a voltage sources in the circuit or there is no tree containing capacitors only and if there is neither an LI-cutset nor a CV-loop with at least one voltage source,

- otherwise, it has index-2.

Proof. For the properly stated leading term we refer to [Mär03]. The index result can be found in [Tis99] and in Theorem 4.3. in [ET00]. $\square$

If the DAE (4.6) has index-2 then the numerically unstable index-2 components are given by currents through voltage sources of CV-loops but also by potentials of inductors and current sources of LI-cutsets, see [Est00, EFM$^+$03]. Fortunately the index-2 variables appear linearly only, see [Est00]. Using perturbation index analysis it has been shown for index-2 Hessenberg systems with linear index-2 variables, [ASW95], and for index-2 circuits, [Tis01], that the numerical difficulties in time integration are moderate, because the differential (index-0) variables are not affected by numerical differentiations.

Next, we add memristor elements to our system. That is, we enlarge our list of basic elements by the memristor. In the MNA framework, we simply add the unknown current

$j_M \in \mathbb{R}^{n_M}$ through the memristors to KCL using the corresponding incidence matrix A_M with n_M the number of memristors. In addition we have to add the characteristic equations for memristors. Starting from the MNA (4.1), we obtain the extended MNA system

$$A_C \frac{d}{dt} q_C \left(A_C^T e, t \right) + A_M \frac{d}{dt} q_M + A_R g_R \left(A_R^T e, t \right) + A_L j_L + A_V j_V + A_I i_s \left(t \right) = 0$$

$$\frac{d}{dt} \phi_M \left(q_M, t \right) - A_M^T e = 0$$

$$\frac{d}{dt} \phi_L \left(j_L, t \right) - A_L^T e = 0 \qquad (4.7)$$

$$A_V^T e - v_s \left(t \right) = 0$$

with the additional unknowns $q_M : \mathcal{I} \to \mathbb{R}^{n_M}$ and characteristic equations $\phi_M \left(q, t \right)$.

For our later investigations we need the following assumptions.

Assumption 4.8 (no short circuit). The matrices A_V and $\begin{bmatrix} A_C & A_R & A_M & A_L & A_V \end{bmatrix}^T$ have full column rank, that is, it exists neither a V-loop nor an I-cutset, see Remark B.14.

Assumption 4.9 (Passivity). The function $\phi_M \left(q, t \right)$ is continuously differentiable with

$$M \left(q, t \right) = \frac{\partial}{\partial q} \phi_M \left(q, t \right)$$

being positive definite.

4.3 Numerical Analysis

In this section we investigate the extended MNA system (4.7) and extend the topological index results for the MNA equations (4.6). The index still depends on simple topological criteria and we see that memristors behave like resistors from the index point of view. Furthermore we provide an approach to calculate a consistent initialization.

The steps are as follows: At first we show that the resulting DAE has a properly stated leading term. Then we develop network topological index-0 conditions. Next we determine the higher index components. With this it is easy to formulate network topological index-1 conditions. Finally we show that the index is always lower or equal two. After the index analysis, we present an approach to compute suitable starting values for the numerical integration.

We suppose that Assumption 4.4 and 4.9 are valid. The extended MNA (4.7) can be written as a DAE given by

$$A \frac{d}{dt} d \left(y, t \right) + b \left(y, t \right) = 0 \qquad (4.8)$$

with unknowns $y = (e, q_M, j_L, j_V)$ and the describing matrix and functions

$$A = \begin{bmatrix} A_C & A_M & 0 & 0 \\ 0 & 0 & I & 0 \\ 0 & 0 & 0 & I \\ 0 & 0 & 0 & 0 \end{bmatrix}, \quad d(y, t) = \begin{pmatrix} q_C\left(A_C^T e, t\right) \\ q_M \\ \phi_M\left(q_M, t\right) \\ \phi_L\left(j_L, t\right) \end{pmatrix}$$

as well as

$$b(y, t) = \begin{pmatrix} A_{R} g_{R}\left(A_R^T e, t\right) + A_L j_L + A_V j_V + A_I i_s(t) \\ -A_M^T e \\ -A_L^T e \\ A_V^T e - v_s(t) \end{pmatrix}.$$

Remark 4.10. In practice, a reformulation of the DAE

$$A\frac{d}{dt}d(y, t) + b(y, t) = 0$$

to a DAE

$$A\frac{d}{dt}\tilde{d}(y, t) + b(y, t) = 0$$

with a properly stated leading term and relations

$$A\frac{d}{dt}d(y, t) = A\frac{d}{dt}\tilde{d}(y, t), \tilde{d}(y, t) = \tilde{P}d(y, t) \text{ and } \ker A = \ker \tilde{P}$$

is not necessary since we are allowed to move the constant projector $\tilde{P}$ from outside into the time derivative and vice versa, see [Mär03]. Moreover with $AD(y, t) = A\frac{\partial}{\partial y}\tilde{d}(y, t)$ the matrix chain is uneffected by the reformulation, too.

Lemma 4.11. Let the Assumption 4.4 and 4.9 be satisfied. Then, the DAE (4.8) has a properly stated leading term, where the constant projector

$$R = \begin{bmatrix} \begin{bmatrix} A_C & A_M \end{bmatrix}^+ \begin{bmatrix} A_C & A_M \end{bmatrix} & \begin{matrix} 0 & 0 \\ 0 & 0 \end{matrix} \\ \begin{matrix} 0 & 0 \\ 0 & 0 \end{matrix} & \begin{matrix} I & 0 \\ 0 & I \end{matrix} \end{bmatrix}$$

realizes the decomposition (2.9).

Proof. The first step is to rewrite the DAE (4.8). For that we choose the projector R with $\ker A = \ker R$, see Lemma A.13, where $\begin{bmatrix} A_C & A_M \end{bmatrix}^+$ denote the Moore-Penrose inverse of $\begin{bmatrix} A_C & A_M \end{bmatrix}$. With that we get

$$\tilde{d}(x, t) = Rd(x, t)$$

and $A \frac{d}{dt} d(y, t) = A \frac{d}{dt} \tilde{d}(y, t)$ holds true. We to show $\operatorname{im} R = \operatorname{im} \frac{\partial}{\partial y} \tilde{d}(y, t)$, see Definition 2.11, that is, it remains to prove

$$\operatorname{im} \begin{bmatrix} A_C & A_M \end{bmatrix}^+ \begin{bmatrix} A_C & A_M \end{bmatrix} = \operatorname{im} \begin{bmatrix} A_C & A_M \end{bmatrix}^+ \begin{bmatrix} A_C C \left(A_C^T e, t \right) A_C^T & A_M \end{bmatrix}$$

That is true since $\operatorname{im} A_C C \left(A_C^T e, t \right) A_C^T = \operatorname{im} A_C$, see Lemma A.3. $\qquad \square$

Remark 4.12. Note, the constant projector

$$\tilde{P} = \begin{bmatrix} A_C^+ A_C & 0 & 0 & 0 \\ 0 & A_M^+ A_M & 0 & 0 \\ 0 & 0 & I & 0 \\ 0 & 0 & 0 & I \end{bmatrix}$$

does not provide a properly stated leading term since $\ker A \not\subseteq \ker \tilde{P}$.

Next we determine the index of the DAE (4.8) by simple topological criteria. We start with the first matrix of the matrix chain, see Definition 2.21, given by

$$G_0(y, t) = \begin{bmatrix} A_C C \left(A_C^T, t \right) A_C^T & A_M & 0 & 0 \\ 0 & M(q_M, t) & 0 & 0 \\ 0 & 0 & L(j_L, t) & 0 \\ 0 & 0 & 0 & 0 \end{bmatrix} \qquad (4.9)$$

with

$$D(y, t) = \begin{bmatrix} C \left(A_C^T e, t \right) A_C^T & 0 & 0 & 0 \\ 0 & I & 0 & 0 \\ 0 & M(q_M, t) & 0 & 0 \\ 0 & 0 & L(j_L, t) & 0 \end{bmatrix}.$$

If the matrix $G_0(y, t)$ is nonsingular, all equations are differential equations, such that the problem is an ODE. This is the case for the following class of circuits.

Theorem 4.13 (index-0). Suppose Assumption 4.4 and 4.9 hold true. The DAE (4.8) has index-0 if and only if there is a tree containing capacitors only and no voltage source.

Proof. We have to check under which conditions the matrix $G_0(y, t)$ is nonsingular. Since $C \left(A_C^T e, t \right)$, $M(q_M, t)$ and $L(j_L, t)$ are positive definite this is the case if and only if the zero rows and columns disappear and $\ker A_C^T = \{0\}$, see Lemma A.3. The null space of A_C^T is trivial if and only if the circuit has a tree containing capacitors only, see Theorem B.11. The zero rows and columns disappear if and only if no voltage sources exist. Using Lemma 2.30 we can conclude that the DAE has index-0. $\qquad \square$

To further continue the matrix chain we need a projector onto $\ker G_0(y, t)$. A possible choice for such a projector is

$$Q_0 = \begin{bmatrix} Q_C & 0 & 0 & 0 \\ 0 & 0 & 0 & 0 \\ 0 & 0 & 0 & 0 \\ 0 & 0 & 0 & I \end{bmatrix}$$

due to Assumption 4.4, where Q_C is constant projectors onto $\ker A_C^T$. For the matrix chain we need the derivative of $b(y, t)$ with respect to the unknowns which is given by

$$
B_0(y, t) = \begin{bmatrix} A_R G\left(A_R^T e, t\right) A_R^T & 0 & A_L & A_V \\ -A_M^T & 0 & 0 & 0 \\ -A_L^T & 0 & 0 & 0 \\ A_V^T & 0 & 0 & 0 \end{bmatrix}.
$$

In addition we calculate

$$
B_0(y, t) Q_0 = \begin{bmatrix} A_R G\left(A_R^T e, t\right) A_R^T Q_C & 0 & 0 & A_V \\ -A_M^T Q_C & 0 & 0 & 0 \\ -A_L^T Q_C & 0 & 0 & 0 \\ A_V^T Q_C & 0 & 0 & 0 \end{bmatrix}.
$$

As already mentioned with regard to the analysis, certain loops and cutsets of edges play a key role. In order to describe different circuit configurations in more detail we will introduce some useful projectors. We denote by

$$
Q_{C-V} \text{ and } Q_{CRMV}
$$

projectors onto

$$
\ker Q_C^T A_V \text{ and } \ker \begin{bmatrix} A_C & A_R & A_M & A_V \end{bmatrix}^T
$$

respectively, see [ET00]. The next lemmata are basically known from [ET00] and slightly extend them to circuits including memristors.

Lemma 4.14 (LI-cutsets). Let a connected circuit be given. The circuit does not contain an LI-cutset if and only if the projector Q_{CRMV} is equal to the zero matrix.

Proof. See Lemma C.2 with $A_{\overline{R}} = \begin{bmatrix} A_R & A_M \end{bmatrix}$ and $A_{\overline{V}} = A_V$. □

Lemma 4.15 (CV-loops). The circuit does not contain a CV-loop with at least one voltage source if and only if the projector Q_{C-V} is equal to the zero matrix.

Proof. See Lemma C.4 with $A_{\overline{V}} = A_V$. □

The next step is the calculation of $\mathcal{N}_0 \cap \mathcal{S}_0(y, t)$. This intersection is crucial for index determination and the consistent initialization as well.

Lemma 4.16. Assume Assumption 4.4 and 4.9 to be satisfied. The index-1 set of the DAE (4.8) can be described by

$$
\mathcal{N}_0 \cap \mathcal{S}_0(y, t) = \{z \in \mathbb{R}^n \mid z_e \in \operatorname{im} Q_{CRMV}, z_{jV} \in \operatorname{im} Q_{C-V}, z_{qM} = 0, z_{jL} = 0\}
$$

for all $(y, t) \in \mathcal{D} \times \mathcal{I}$.

Proof. For calculating the index-1 set we make use of Remark 2.27. For a suitable description we make use of a projector along $\operatorname{im} G_0(y,t)$. We are given one by

$$
W_0(y,t) = \begin{bmatrix} Q_C^\mathsf{T} & -Q_C^\mathsf{T} A_M M(q_M,t)^{-1} & 0 & 0 \\ 0 & 0 & 0 & 0 \\ 0 & 0 & 0 & 0 \\ 0 & 0 & 0 & I \end{bmatrix},
$$

see Lemma C.13, and we get

$$
W_0(y,t)\,B_0(y,t)\,Q_0 = \begin{bmatrix} Q_C^\mathsf{T} G(e,q_M,t)\,Q_C & 0 & 0 & Q_C^\mathsf{T} A_V \\ 0 & 0 & 0 & 0 \\ 0 & 0 & 0 & 0 \\ A_V^\mathsf{T} Q_C & 0 & 0 & 0 \end{bmatrix}
$$

with $G(e,q_M,t) = \left(A_R G\left(A_R^\mathsf{T} e, t\right) A_R^\mathsf{T} + A_M M(q_M,t)^{-1} A_M^\mathsf{T} \right)$.

Let be $z \in \operatorname{im} Q_0 \cap \ker W_0(y,t)\,B_0(y,t)\,Q_0$. That is true if and only if

$$Q_C z_e = z_e \tag{4.10}$$
$$z_{q_M} = 0$$
$$z_{j_L} = 0$$
$$Q_C^\mathsf{T} \left(A_R G\left(A_R^\mathsf{T} e, t\right) A_R^\mathsf{T} + A_M M(q_M,t)^{-1} A_M^\mathsf{T} \right) Q_C z_e + Q_C^\mathsf{T} A_V z_{j_V} = 0 \tag{4.11}$$
$$A_V^\mathsf{T} Q_C z_e = 0 \tag{4.12}$$

hold true, using Assumption 4.4 and 4.9. Left-multiply (4.11) by z_e^T and using (4.12) leads to $Q_C z_e \in \ker \begin{bmatrix} A_R & A_M \end{bmatrix}^\mathsf{T}$, see Lemma A.3. We obtain $z_e \in \operatorname{im} Q_{CRMV}$ in combination with (4.10) and (4.12). Thus (4.11) leads to

$$Q_C^\mathsf{T} A_V z_{j_V} = 0 \text{ and } z_{j_V} \in \operatorname{im} Q_{C-V}.$$

We get $z \in \mathcal{N}_0 \cap \mathcal{S}_0(y,t)$ if and only if

$$z_e \in \operatorname{im} Q_{CRMV}$$
$$z_{q_M} = 0$$
$$z_{j_L} = 0$$
$$z_{j_V} \in \operatorname{im} Q_{C-V}$$

holds true. □

Remark 4.17. It is possible to choose a constant projector along $\operatorname{im} G_0(y,t)$, since $\operatorname{im} A = \operatorname{im} G_0(y,t)$, see Lemma 2.12. Nonetheless it is more convenient to make use of the given non-constant projector $W_0(y,t)$ to prove Lemma 4.16.

With the characterization of $\mathcal{N}_0 \cap \mathcal{S}_0(y,t)$ we are able to provide network topological index-1 conditions. We show that, from the index point of view, the memristors behave like resistors.

Theorem 4.18 (index-1). Let Assumption 4.4 and 4.9 to be true. The DAE (4.8) has index-1 if and only if there is at least a voltage sources in the circuit or there is no tree containing capacitors only and if there is neither an LI-cutset nor a CV-loop with at least one voltage source.

Proof. We make use of the representation of $\mathcal{N}_0 \cap \mathcal{S}_0 (y, t)$ as proposed in Lemma 4.16. The intersection $\mathcal{N}_0 \cap \mathcal{S}_0 (y, t)$ is trivial if and only if $Q_{CRMV} = 0$ and $Q_{C-V} = 0$. This is equivalent the condition that to the circuit containing neither LI-cutsets nor CV-loops, see Lemma 4.14 and 4.15. Using Definition 2.23 we get the DAE (4.8) to be of index-1. $\qquad\square$

The DAE (4.8) can be of index-2 also, but higher index problems can be avoided as we will see in the next theorem. We will see that LI-cutsets and CV-loops are the only critical circuit configurations.

Obviously the dimension of $\mathcal{N}_0 \cap \mathcal{S}_0 (x, t)$ is constant, which is important for the index-2 case.

Theorem 4.19 (index-2). Let Assumption 4.4, 4.8 and 4.9 hold true. The DAE (4.8) has index-2 if and only if there is an LI-cutset or a CV-loop with at least one voltage source.

Proof. At first we need

$$
G_1(y, t) = \begin{bmatrix}
A_C C \left(A_C^\top e, t\right) A_C^\top + A_R G \left(A_R^\top e, t\right) A_R^\top Q_C & A_M & 0 & A_V \\
-A_M^\top Q_C & M(q_M, t) & 0 & 0 \\
-A_L^\top Q_C & 0 & L(j_L, t) & 0 \\
A_V^\top Q_C & 0 & 0 & 0
\end{bmatrix} \quad (4.13)
$$

in order to proceed the matrix chain. For the characterization of the index-2 set we introduce a projector along $\operatorname{im} G_1(y, t)$, see Remark 2.28. We are given one by

$$
W_1 = \begin{bmatrix}
Q_{CRMV}^\top & 0 & 0 & 0 \\
0 & 0 & 0 & 0 \\
0 & 0 & 0 & 0 \\
0 & 0 & 0 & Q_{C-V}^\top
\end{bmatrix},
$$

see Lemma C.14. Next we take into account

$$
P_0 = \begin{bmatrix}
P_C & 0 & 0 & 0 \\
0 & I & 0 & 0 \\
0 & 0 & I & 0 \\
0 & 0 & 0 & 0
\end{bmatrix}, \quad
B_0(y, t) P_0 = \begin{bmatrix}
A_R G \left(A_R^\top e, t\right) A_R^\top P_C & 0 & A_L & 0 \\
-A_M^\top P_C & 0 & 0 & 0 \\
-A_L^\top P_C & 0 & 0 & 0 \\
A_V^\top P_C & 0 & 0 & 0
\end{bmatrix},
$$

where P_0 is the complementary projector to Q_0 and

$$
W_1 B_0(y, t) P_0 = \begin{bmatrix}
0 & 0 & Q_{CRMV}^\top A_L & 0 \\
0 & 0 & 0 & 0 \\
0 & 0 & 0 & 0 \\
Q_{C-V}^\top A_V^\top P_C & 0 & 0 & 0
\end{bmatrix}.
$$

Let be $z \in \ker G_1(y,t) \cap \ker W_1 B_0(y,t) P_0$. That is true if and only if the conditions

$$Q_{CRMV}^T A_L z_{j_L} = 0 \tag{4.14}$$

$$Q_{C-V}^T A_V^T P_C z_e = 0 \tag{4.15}$$

$$Q_C z_e \in \operatorname{im} Q_{CRMV} \tag{4.16}$$

$$z_{j_V} \in \operatorname{im} Q_{C-V} \tag{4.17}$$

$$z_{q_M} = 0$$

$$P_C z_e = -H_C \left(A_C^T e, t\right)^{-1} A_V Q_{C-V} z_{j_V} \tag{4.18}$$

$$L\left(j_L, t\right)^{-1} A_L^T Q_C z_e = z_{j_L} \tag{4.19}$$

are fulfilled, using

$$\mathcal{N}_1(y,t) = \left\{ z \in \mathbb{R}^n \,|\, Q_C z_e \in \operatorname{im} Q_{CRMV}, z_{j_V} \in \operatorname{im} Q_{C-V}, z_{q_M} = 0, \right.$$
$$\left. P_C z_e = -H_C \left(A_C^T e, t\right)^{-1} A_V Q_{C-V} z_{j_V}, L\left(j_L, t\right)^{-1} A_L^T Q_C z_e = z_{j_L} \right\},$$

see Lemma C.15, where $H_C\left(A_C^T e, t\right) = A_C C\left(A_C^T e, t\right) A_C^T + Q_C^T Q_C$ is positive definite, see Lemma A.10, using Assumption 4.4 and 4.9. From (4.16) we deduce $Q_C z_e = Q_{CRMV} Q_C z_e$. Left-multiplying (4.14) by $z_e^T Q_C^T$ and inserting of (4.19) yields

$$z_e^T Q_C^T L\left(j_L, t\right)^{-1} A_L^T Q_C z_e = 0 \text{ and } Q_C z_e \in \ker A_L^T,$$

see Lemma A.3. Hence $Q_C z_e \in \ker \begin{bmatrix} A_C & A_V & A_R & A_M & A_L \end{bmatrix}^T$ and we conclude that $Q_C z_e = 0$, since I-cutsets are prohibited. Consequently (4.19) leads to $z_{j_L} = 0$. Inserting (4.18) in (4.15) and using (4.17) we obtain $A_V z_{j_V} = 0$. Thus $z_{j_V} = 0$ due to V-loops are forbidden. From (4.18) we get $z_e = 0$ and we result in $z = 0$, see Definition 2.23. $\qquad \square$

To start the integration of the DAE (4.8) we need a consistent initialization. In case of index-1 we make direct use of Theorem 2.51. In case of index-2 we apply Theorem 2.58. For this we need to check the requirements.

Assumption 4.20. For the DAE (4.8) exist the continuous partial derivatives $\frac{\partial}{\partial t} d(y,t)$ and $\frac{\partial}{\partial t} W_1 b(y,t)$ for all $(y,t) \in \mathcal{D} \times \mathcal{I}$.

These assumptions are not a restriction since, if a solution exists, then $\frac{\partial}{\partial t} d(y,t)$ exists and is continuous. Moreover $W_1 b(y,t)$ describes exactly the hidden constraints and hence $\frac{\partial}{\partial t} W_1 b(y,t)$ needs to exists and to be continuous to have a solution of the problem. In addition, the DAE (4.8) has a constant matrix A and there are the constant projectors Q_0 and W_1. It remains to show that the index-2 variables enter linearly only.

Lemma 4.21. The relation

$$g_R\left(A_R^T e, t\right) = g_R\left(A_R^T P_{CRMV} e, t\right)$$

holds true for all $(e,t) \in \mathbb{R}^{n_N} \times \mathcal{I}$.

Proof. We apply the mean value theorem. We get

$$g_R\left(A_R^\top e, t\right) - g_R\left(A_R^\top P_{\mathrm{CRMV}} e, t\right) = \int_0^1 G\left(sA_R^\top e + (1-s) A_R^\top P_{\mathrm{CRMV}} e, t\right) A_R^\top Q_{\mathrm{CRMV}} e ds$$
$$= 0.$$

$\square$

Lemma 4.22. Let Assumption 4.4 and 4.9 be fulfilled. The index-2 variables enter the DAE (4.8) linearly only.

Proof. From Lemma 4.16 we easily obtain a constant projector T onto $\mathcal{N}_0 \cap \mathcal{S}_0\left(y, t\right)$ given by

$$T = \begin{bmatrix} Q_{\mathrm{CRMV}} & 0 & 0 & 0 \\ 0 & 0 & 0 & 0 \\ 0 & 0 & 0 & 0 \\ 0 & 0 & 0 & Q_{\mathrm{C-V}} \end{bmatrix}$$

and the complementary projector U reads

$$U = \begin{bmatrix} P_{\mathrm{CRMV}} & 0 & 0 & 0 \\ 0 & I & 0 & 0 \\ 0 & 0 & I & 0 \\ 0 & 0 & 0 & P_{\mathrm{C-V}} \end{bmatrix}.$$

The unknowns are divided into

$$y = Ty + Uy = \begin{pmatrix} Q_{\mathrm{CRMV}} e \\ 0 \\ 0 \\ Q_{\mathrm{C-v}} j v \end{pmatrix} + \begin{pmatrix} P_{\mathrm{CRMV}} e \\ q_M \\ j_L \\ P_{\mathrm{C-v}} j v \end{pmatrix}.$$

Now we can write $b\left(y, t\right) = b\left(Uy, t\right) + BTy$ with

$$B = \begin{bmatrix} 0 & 0 & 0 & A_V \\ 0 & 0 & 0 & 0 \\ -A_L^\top & 0 & 0 & 0 \\ 0 & 0 & 0 & 0 \end{bmatrix},$$

using Lemma 4.21. The relation $d\left(y, t\right) = d\left(Uy, t\right)$ is obvious by Lemma 2.54. $\square$

The DAE (4.8) fulfills all requirements to apply Theorem 2.58 in case of index-2. But we still need an operating point when we want to integrate it numerically. Since Theorem 2.58 is applicable to the DAE an operating point is sufficient to start the numerical integration, see Lemma 2.61.

The computation of an operating point and DC (direct current) solution is very well established area, see [CL75, CDK87, DK84, SV93], and is usually the first step in the simulation of ciruits. Common approaches are, among others, homotopy methods and source ramping, which are used in SPICE and TITAN, [Vla94, SW96, Dau10].

Let be $z^0 = \left(z_C^0, z_M^0, z_{\phi_M}^0, z_L^0\right)$, $y^0 = \left(e^0, q_M^0, j_L^0, j_V^0\right)$ and $t_0 \in \mathcal{I}$. We choose (z_C^0, z_M^0) and (q_M^0, j_L^0) such that

$$A_C z_C^0 + A_M z_M^0 + A_L j_L^0 + A_I i_s\left(t_0\right) \in \operatorname{im}\begin{bmatrix} A_R & A_V \end{bmatrix}. \tag{4.20}$$

Definition 4.23 (RV-*path*). A path is called RV-path if and only if the path containing resistors and voltage sources only.

Remark 4.24. The condition (4.20) can be fulfilled if we choose all currents through capacitors, memristors and inductors to be zero. For capacitors, memristors and inductors, where the elements contacts are connected by a RV-path we can choose freely the currents and charges through the elements, respectively. For simplicity we assume the currents through current sources to be zero if the contacts are not connected by a RV-path, otherwise we apply a source ramping, see [Vla94].

Next we determine (e^0, j_V^0). For this we need a solution of the nonlinear system:

$$\begin{aligned} A_R g_R\left(A_R^\top e^0, t_0\right) + A_V j_V^0 &= -A_C z_C^0 - A_M z_M^0 - A_L j_L^0 - A_I i_s\left(t_0\right) \\ A_V^\top e^0 &= v_s\left(t_0\right) \end{aligned} \tag{4.21}$$

The Jacobian of the nonlinear system (4.21) has the form

$$J_{ev}\left(e^0, j_V^0\right) = \begin{bmatrix} A_R G\left(A_R^\top e^0, t_0\right) A_R^\top & A_V \\ A_V^\top & 0 \end{bmatrix}.$$

Let be $v \in \ker J_{ev}\left(e^0, j_V^0\right)$. Then we get:

$$A_R G\left(A_R^\top e^0, t_0\right) A_R^\top v_1 + A_V v_2 = 0 \tag{4.22}$$

$$A_V^\top v_1 = 0 \tag{4.23}$$

Left multiplying (4.22) by $v_1^\top$ and using (4.23) leads to $v_1 \in \ker\begin{bmatrix} A_R & A_V \end{bmatrix}^\top$ and hence

$$\ker J_{ev}\left(e^0, j_V^0\right) = \ker\begin{bmatrix} A_R & A_V \end{bmatrix}^\top \times \{0\}. \tag{4.24}$$

In analogy we show $\ker J_{ev}\left(e^0, j_V^0\right) = \ker J_{ev}\left(e^0, j_V^0\right)^\top$.

Approach 4.25. To solve the system (4.21) we suggest two possible ways:

(i) Assume $\begin{bmatrix} A_R & A_V \end{bmatrix}$ to have full row rank. In terms of network configurations that means there is a tree containing voltage sources and resistors only, that is, the index-2 configurations are CV-loops only. That is, we can choose (z_C^0, z_M^0) and

(q_M^0, j_L^0) arbitrarily. Let be $v \in \ker J_{ev}(e^0, j_V^0)$. Then we deduce from (4.24) the relation $v_1 \in \ker \begin{bmatrix} A_R & A_V \end{bmatrix}^T$ and $v_1 = 0$ due to $\begin{bmatrix} A_R & A_V \end{bmatrix}$ has full row rank. Furthermore we get $v_2 = 0$ by Assumption 4.8, that is, V-loops are forbidden. With that we conclude $J_{ev}(e^0, j_V^0)$ to be is nonsingular. Hence we obtain a unique solution for (e^0, j_V^0) by solving the nonlinear system (4.21) using, for example, Newton's method.

(ii) We apply Theorem 2.59 to (4.21). Then

$$\left(e^0, j_V^0\right) = B_P v + u$$

with $u \in \ker B_P B_P^T$ being arbitrarily and $v \in \mathbb{R}^k$ is the unique solution of

$$B_P^T f \left(B_P v, t_0\right) = 0,$$

where

$$f\left((e, j_V), t\right) = \begin{pmatrix} A_R g_R \left(A_R^T e, t\right) + A_V j_V + A_C z_C^0 + A_M z_M^0 + A_L j_L^0 + A_I i_s(t) \\ A_V^T e - v_s(t) \end{pmatrix}.$$

After applying Approach 4.25 we compute the missing parts by

$$z_{\phi_M}^0 = A_M^T e^0 \text{ and } z_L^0 = A_L^T e^0.$$

Remark 4.26. Due to the structure of the DAE (4.8) we obtain a locally unique solution through every consistent inital value and the perturbation index to be not greater than two, see Theorem 2.33, 2.46, 2.49 and 2.50.

4.4 Summary

In this chapter we have introduced the modified nodal analysis to model circuits containing the basic elements including memristors formulated as a differential-algebraic equation (4.8) with a properly stated leading term.

We extended the well-known topological index conditions of [Tis99, ET00] for the modified nodal analysis to circuits including memristors (Theorem 4.13, 4.18 and 4.19) and showed that the index does not exceed two. We conclude that, from index point of view, the memristors behave like resistors. Moreover. we have shown perturbation and solvability results for the modified nodal analysis including memristors (Remark 4.26) and the perturbation index does not exceed two.

We presented two approaches (Approach 4.25) for the calculation of an operating point. Based on the linearity of the index-2 components (Lemma 4.21) the calculation of a consistent initialization is possible by correcting an operating point by solving a linear system. Due to the structure it is sufficient to start the numerical integration with the implicit Euler using an operating point to obtain a consistent initialization after the first time step.

5 Coupled Electromagnetic Field/Circuit Models

Usually in a technology computer aided design environment devices exhibiting multi-physical effects such as electromagnetic or semiconductor devices are simplified and the devices are modeled by an equivalent circuits.

The rapid developments in chip technology lead the devices being ever more minimized and higher frequencies evoking effects that no longer can be reproduced by an equivalent circuit in an appropriate manner. One reason is that the performance of the devices is significantly influenced by the surrounding circuitry such as, for example, heating or inductive coupling. This requires additional iterations during the circuit design for the extraction and generation of equivalent circuit parameters for the different time steps in simulation. Today, the equivalent circuits such as the BSIM6 transistor models (University of California Berkeley Device Group) depend on up to hundreds of parameters. Most of these parameters do not have a direct physical interpretation, see [DF06], and their calibration is a time consuming and challenging task.

To meet future demands in circuit design it is recommended to combine circuit simulation directly with device simulation. While most elements are modeled by equivalent circuits we simulate a particular device with a refined model to meet the contradicting requirements of correct physical behavior of the circuit and reasonably small simulation time.

In engineering it is a common task to couple circuit and device simulation, see [Tis04] and references therein. But for mathematics it is a young research area. Several index results for circuits with various distributed elements leading to differential-algebraic equations have beed proposed during the last years: lossy transmission lines [Gün01], heating [Bar04, Cul09] and semiconductors [Tis04, ST05, Sel06, Bod07, BST12]. More theoretical results concerning solvability of abstract differential algebraic equations, that is, differential-algebraic equations on infinite dimensional Banach spaces, are presented in [Tis04, Rei06, Mat12] with their results being also applicable to circuits including partial differential equation models.

We investigate coupled electromagnetic device/circuit models with spatially resolved electromagnetic devices. The electromagnetic devices are described by Maxwell's equations in a potential formulation and spatially discretized by the finite integration technique. Coupled magnetoquasistatic device/circuit models are investigated in [HM76, KMST93, DHW04, DW04] and index results are presented in [Tsu02, Ben06, BBS11, Sch11] using certain conductor models like stranded and solid conductors. The magnetoquasistatic assumption leads to the eddy-current problem for the device. In [Tsu02, Ben06] index-1 circuit configurations are investigated while [BBS11, Sch11] take general circuit configurations into account by extending the topological index conditions for the modified nodal analysis given in [Tis99, ET00]. Our index analysis for coupled electromagnetic device/circuit models does not cover a special class of conductor models and we do not suppose that the magnetoquasistatic assumption holds. It turns out that the index of the coupled system depends on the chosen gauge condition. For the coupled electromagnetic device/circuit model using Lorenz gauge we extend the topological index conditions for the modified nodal analysis. The Coulomb gauge always results to an index-2 differential-algebraic equation.

This chapter is devoted to the index analysis of coupled systems. First, we introduce the terminology of coupled system and point out why an index analysis is necessary. Next, we introduce the coupled system consisting of circuits refined by spatially resolved electromagnetic devices modeled by the modified nodal analysis and spatial discretized Maxwell's equations formulated as a differential-algebraic equation with a properly stated leading term. We generalize the topological index criteria for the modified nodal analysis to the coupled system. In addition, we present an approach to calculate a consistent initialization.

5.1 Simulation of Coupled Systems

Mathematical models for coupled systems are characterized by their decomposition into different subsystems described by differential equations in space and time. These subsystems may arise through refined modeling. The interdependencies are named *coupling conditions* and describe the mutal impact of the subsystems. There are two major approaches for the time integration of coupled systems:

- *cosimulation*: The subsystems are solved sequentially or in parallel. The information interchange is restricted to particular time points. All subsystems may be solved on their own time scale (multirate) with tailor-made methods (multimethod). We call cosimulation systems *weakly coupled*. Cosimulation requires more detailed analysis of the system formulation and the coupling conditions.

- *monolithic*: All subsystems are combined into one single system of equations and solved simultaneously. Every subsystem has all system information at every time point. All subsystems must be solved on the same time scale using the same methods. We call monolithic systems *strongly coupled*.

In this thesis monolithic systems of DAEs are investigated. We would like to stress that it is not sufficient to determine the index of the different subsystems to deduce the index

of the coupled system as shown by the following examples. The index of the coupled
system depends on the structure of the subsystems as well as on the structure of the
coupling conditions.

Example 5.1. We show that coupling of an index-1 and index-2 DAE can result in a
monolithic index-1 or index-2 DAE. Let us consider the following system

$$\frac{d}{dt}x_1 = x_2 \qquad\qquad \frac{d}{dt}y_1 = y_2$$
$$x_2 = u(t) \qquad\qquad y_1 = v(t)$$

where $u(t)$ and $v(t)$ are given inputs and the subsystems consist of an index-1 and an
index-2 DAE. The coupling conditions are given by

(i)

$$u = y_1 + y_2 \text{ and } v = x_1 + x_2$$

(ii)

$$u = y_1 \text{ and } v = x_1$$

such that the monolithic system using (i) leads to an index-1 and using (ii) leads to an
index-2 DAE.

Example 5.2. We show that coupling two index-1 DAEs can result in a monolithic
index-2 DAE and vice versa. Let us consider the following systems

(i)

$$\frac{d}{dt}x_1 = x_2 \qquad\qquad \frac{d}{dt}y_1 = y_2$$
$$x_1 + x_2 = u(t) \qquad\qquad y_1 + y_2 = v(t)$$

(ii)

$$\frac{d}{dt}x_1 = x_2 \qquad\qquad \frac{d}{dt}y_1 = y_2$$
$$x_1 = u(t) \qquad\qquad y_1 = v(t)$$

where $u(t)$ and $v(t)$ are given inputs and the subsystem (i) consists of an index-1 and
(ii) of an index-2 DAE. The coupling conditions are given by

$$u = y_2 \text{ and } v = x_2$$

such that the monolithic system (i) leads to an index-2 and (ii) leads to an index-1 DAE.

5.2 Electromagnetic Field/Circuit Model

In this section we investigate circuits refined by spatially resolved EM devices modeled by the MNA and ME. The MNA describes the non-critical circuit parts for which a modeling using basic elements only is sufficient. Critical circuit parts for which the MNA approach is insufficient to describe the EM device behavior are modeled by ME directly. For simplicity we assume that only one critical EM device is given.

We have to include the EM device into the MNA framework. For this charge conservation is essential which is given by the additional mass contact. We suppose that the EM device has n_E disjoint conductive contacts and each contact of the device is joined to a node of the circuit. In addition we suppose that the mass contact is connected to the mass node. The contacts of the EM device joined to the same node of the electrical circuit define a terminal, see [Tis04, Bod07]. Let $n_T + 1$ be the number of terminals of the EM device and n_N be the number of circuit's nodes except the mass node. We define the following (reduced) incidence matrix $A_E \in \{-1, 0, 1\}^{n_N \times n_T}$ by

$$
(A_E)_{ij} = \begin{cases} 1 & \text{if terminal } j \text{ is joined to node } i, \\ -1 & \text{if the reference terminal is attached to node } i, \\ 0 & \text{else.} \end{cases}
$$

The coupling of the EM device to the circuit is established by the applied node potentials at the EM device conductive contacts and the currents through it. For this we need to add the EM device to our list of elements. In the MNA framework we simply add the current through the EM device to the KCL using the corresponding incidence matrix A_E. In addition we add the MGE (3.60) for the EM device to the MNA (4.6), where the Dirichlet boundary conditions for the scalar potentials are described by $e = A_E^T e$. That is, we apply the potential difference to the mass node at the conductive contacts. That is possible since the scalar potentials are determined up to a constant. The coupled EM device/circuit system with gauge condition reads

$$
A_C \frac{\mathrm{d}}{\mathrm{d}t} q_C \left(A_C^T e, t \right) + A_R g_R \left(A_R^T e, t \right) + A_L j_L + A_V j_V + A_E j_E + A_I i_s \left(t \right) = 0
$$

$$
\frac{\mathrm{d}}{\mathrm{d}t} \phi_L \left(j_L, t \right) - A_L^T e = 0
$$

$$
A_V^T e - v_s \left(t \right) = 0
$$

$$
j_E - \Lambda_u^T K_\nu^u (\widehat{a}_u) \widehat{a}_u = 0
$$

$$
\vartheta \widetilde{S}_u M_\varepsilon^u G_u \frac{\mathrm{d}}{\mathrm{d}t} \phi_u + \widetilde{S}_u M_{\widetilde{\nu}}^u \widehat{a}_u = 0 \tag{5.1}
$$

$$
-M_\varepsilon^u \Lambda_u A_E^T \frac{\mathrm{d}}{\mathrm{d}t} e + M_\varepsilon^u G_u \frac{\mathrm{d}}{\mathrm{d}t} \phi_u + M_\varepsilon^u \frac{\mathrm{d}}{\mathrm{d}t} \widehat{\pi}_u - M_\sigma^u \Lambda_u A_E^T e
$$
$$
+ M_\sigma^u G_u \phi_u + K_\nu^u (\widehat{a}_u) \widehat{a}_u + M_\sigma^u \widehat{\pi}_u = 0
$$

$$
\frac{\mathrm{d}}{\mathrm{d}t} \widehat{a}_u - \widehat{\pi}_u = 0
$$

in time $t \in \mathcal{I}$, $\mathcal{I} = [t_0, \ T] \subset \mathbb{R}$, see Remark 3.33.

For our analysis of the coupled system (5.1) certain loops and cutsets play a key role.

Assumption 5.3 (no short circuit)**.** The matrices A_V and $\begin{bmatrix} A_C & A_R & A_L & A_E & A_V \end{bmatrix}^\mathsf{T}$ have full column rank, that is, it exists neither a V-loop nor an I-cutset, see Remark B.14.

Definition 5.4 (LEI-cutset)**.** A cutset is called LEI-cutset if and only if the cutset contains only inductors, EM devices and current sources.

In order to describe different circuit configurations in more detail we will introduce some useful projectors. We denote by

$$Q_C, \ Q_{C-V} \text{ and } Q_{CRV}$$

projectors onto

$$\ker A_C^\mathsf{T}, \ \ker Q_C^\mathsf{T} A_V \text{ and } \ker \begin{bmatrix} A_C & A_R & A_V \end{bmatrix}^\mathsf{T}$$

respectively, see [ET00]. The next lemmata are basically known from [ET00] and we slightly extend them to circuits including EM devices.

Lemma 5.5 (LEI-cutsets)**.** Let a connected circuit be given. The circuit does not contain an LEI-cutset if and only if the projector Q_{CRV} is equal to the zero matrix.

Proof. See Lemma C.2 with $A_{\overline{R}} = A_R$ and $A_{\overline{V}} = A_V$. □

Lemma 5.6 (CV-loops)**.** The circuit does not contain a CV-loop with at least one voltage source if and only if the projector Q_{C-V} is equal to the zero matrix.

Proof. See Lemma C.4 with $A_{\overline{V}} = A_V$. □

5.3 Numerical Analysis

In this section we investigate the coupled system (5.1) using Coulomb and Lorenz gauge. For both systems we extend the topological index results for the MNA (4.6), see [Tis99, ET00]. The index depends still on simple topological criteria and we see that an EM device using Lorenz gauge, from the index point of view, behaves like an inductor. Furthermore we provide an approach to calculate a consistent initialization.

We suppose that Assumption 4.4, 3.31 and Property 3.32 are fulfilled.

The steps are as follows: First we show that the resulting DAEs have a properly stated leading term. Then we develop network topological index-0 conditions. Next we determine the higher index components. With this it is easy to formulate network topological index-1 conditions. Finally we show that the index is always lower or equal two. After the index analysis, we present an approach to compute suitable starting values for the numerical integration.

5.3.1 Field/Circuit System using Coulomb Gauge

The coupled system (5.1) using Coulomb gauge, that is, $\vartheta = 0$, can be formulated as a DAE given by

$$A\frac{d}{dt}d(y,t) + b(y,t) = 0 \qquad (5.2)$$

with unknowns $y = (e, j_L, j_V, j_E, \phi_u, \widehat{a}_u, \widehat{\pi}_u)$ and the describing matrix and functions

$$A = \begin{bmatrix} A_C & 0 & 0 & 0 \\ 0 & I & 0 & 0 \\ 0 & 0 & 0 & 0 \\ 0 & 0 & 0 & 0 \\ 0 & 0 & 0 & 0 \\ 0 & 0 & 0 & M_\varepsilon^u \\ 0 & 0 & I & 0 \end{bmatrix}, \quad d(y,t) = \begin{pmatrix} q_C(A_C^\top e, t) \\ \phi_L(j_L, t) \\ \widehat{a}_u \\ -\Lambda_u A_E^\top e + G_u \phi_u + \widehat{\pi}_u \end{pmatrix}$$

as well as

$$b(y,t) = \begin{pmatrix} A_R g_R (A_R^\top e, t) + A_L j_L + A_V j_V + A_E j_E + A_I i_s(t) \\ -A_L^\top e \\ A_V^\top e - v_s(t) \\ j_E - \Lambda_u^\top K_\nu^u(\widehat{a}_u)\widehat{a}_u \\ \widetilde{S}_u M_\nu^u \widehat{a}_u \\ -M_\sigma^u \Lambda_u A_E^\top e + M_\sigma^u G_u \phi_u + K_\nu^u(\widehat{a}_u)\widehat{a}_u + M_\sigma^u \widehat{\pi}_u \\ -\widehat{\pi}_u \end{pmatrix}.$$

First, we show that the DAE has a properly stated leading term.

Lemma 5.7. Let Assumption 4.4 and 3.31 be fulfilled. Then, the DAE (5.2) has a properly stated leading term where the constant projector

$$R = \begin{bmatrix} A_C^+ A_C & 0 \\ 0 & I \end{bmatrix}$$

realizes the decomposition (2.9).

Proof. The first step is to rewrite the DAE (5.2). For that we choose a projector

$$\widetilde{P} = \begin{bmatrix} A_C^+ A_C & 0 \\ 0 & I \end{bmatrix}$$

with $A = A\widetilde{P}$, where A_C^+ denote the Moore-Penrose inverse of A_C. With this we get

$$\widetilde{d}(x,t) = \widetilde{P}d(x,t)$$

and $A\frac{d}{dt}d\,(y,t) = A\frac{d}{dt}\tilde{d}\,(y,t)$ holds true. We denote $\tilde{D}\,(y,t) = \frac{\partial}{\partial y}\tilde{d}\,(y,t)$ given by

$$\tilde{D}\,(y,t) = \begin{bmatrix} A_C^+ A_C C\left(A_C^\top e,t\right) A_C^\top & 0 & 0 & 0 & 0 & 0 \\ 0 & L\,(j_L,t) & 0 & 0 & 0 & 0 \\ 0 & 0 & 0 & 0 & I & 0 \\ -\Lambda_u A_E^\top & 0 & 0 & 0 & G_u & 0 & I \end{bmatrix}$$

We get

$$\ker A = \ker A_C \times \{0\}$$

and

$$\operatorname{im}\tilde{D}\,(y,t) = \operatorname{im}A_C^+ A_C C\left(A_C^\top e,t\right) A_C^\top \times \mathbb{R}^{n_L+n_a+n_\pi},$$

using Assumption 4.4 and 3.31. Applying Lemma A.1, A.19 and A.21 we obtain

$$\operatorname{im}A_C^+ A_C C\left(A_C^\top e,t\right) A_C^\top = \operatorname{im}A_C^+ A_C = \operatorname{im}A_C^\top.$$

Hence we can choose the projector $R = \tilde{P}$, see Lemma A.13. $\qquad\square$

Notice that the projector $\tilde{P}$ in Lemma 5.7 is not needed for practical computations, see Remark 4.10.

We follow the matrix chain concept, see Definition 2.21. For that we need the matrix

$$G_0\,(y,t) = \begin{bmatrix} A_C C\left(A_C^\top e,t\right) A_C^\top & 0 & 0 & 0 & 0 & 0 & 0 \\ 0 & L\,(j_L,t) & 0 & 0 & 0 & 0 & 0 \\ 0 & 0 & 0 & 0 & 0 & 0 & 0 \\ 0 & 0 & 0 & 0 & 0 & 0 & 0 \\ 0 & 0 & 0 & 0 & 0 & 0 & 0 \\ -M_\varepsilon^u \Lambda_u A_E^\top & 0 & 0 & 0 & M_\varepsilon^u G_u & 0 & M_\varepsilon^u \\ 0 & 0 & 0 & 0 & 0 & I & 0 \end{bmatrix} \qquad (5.3)$$

with

$$D\,(y,t) = \begin{bmatrix} C\left(A_C^\top e,t\right) A_C^\top & 0 & 0 & 0 & 0 & 0 & 0 \\ 0 & L\,(j_L,t) & 0 & 0 & 0 & 0 & 0 \\ 0 & 0 & 0 & 0 & I & 0 \\ -\Lambda_u A_E^\top & 0 & 0 & 0 & G_u & 0 & I \end{bmatrix}.$$

To obtain an index-0 DAE we need to check under which conditions the matrix $G_0\,(y,t)$ is nonsingular. If the matrix $G_0\,(y,t)$ is nonsingular all equations are differential equations, such that the problem is an ODE. This is the case for the following class of circuits.

Theorem 5.8 (index-0). Let Assumption 4.4 be fulfilled. The DAE (5.2) has index-0 if and only if the circuit does not contain voltage sources and EM device and if there is a tree containing capacitors only.

Proof. Following the proof of Theorem 4.13, the remaining zero rows and columns disappear if and only if there are no EM device. $\qquad \square$

To further continue the matrix chain we need a projector onto $\ker G_0(y,t)$. Let be $z \in \ker G_0(y,t)$. That is true if and only if

$$z_e \in \operatorname{im} Q_C$$
$$z_{j_L} = 0$$
$$z_{\widehat{\pi}_u} = \Lambda_u A_E^\top Q_C z_e - G_u z_{\phi_u}$$
$$z_{\widehat{a}_u} = 0$$

hold true, due to Lemma A.3, Assumption 4.4 and 3.31. We can choose a constant projector onto $\ker G_0(y,t)$ by

$$Q_0 = \begin{bmatrix} Q_C & 0 & 0 & 0 & 0 & 0 & 0 \\ 0 & 0 & 0 & 0 & 0 & 0 & 0 \\ 0 & 0 & I & 0 & 0 & 0 & 0 \\ 0 & 0 & 0 & I & 0 & 0 & 0 \\ 0 & 0 & 0 & 0 & I & 0 & 0 \\ 0 & 0 & 0 & 0 & 0 & 0 & 0 \\ \Lambda_u A_E^\top Q_C & 0 & 0 & 0 & -G_u & 0 & 0 \end{bmatrix}.$$

For the matrix chain we need the derivative of $b(y,t)$ with respect to the unknowns which is given by

$$B_0(y,t) = \begin{bmatrix} A_R G\left(A_R^\top e, t\right) A_R^\top & A_L & A_V & A_E & 0 & 0 & 0 \\ -A_L^\top & 0 & 0 & 0 & 0 & 0 & 0 \\ A_V^\top & 0 & 0 & 0 & 0 & 0 & 0 \\ 0 & 0 & 0 & I & 0 & -\Lambda_u^\top K_{\nu,d}^u(\widehat{a}_u) & 0 \\ 0 & 0 & 0 & 0 & 0 & \widetilde{S}_u M_{\overline{\nu}}^u & 0 \\ -M_\sigma^u \Lambda_u A_E^\top & 0 & 0 & 0 & M_\sigma^u G_u & K_{\nu,d}^u(\widehat{a}_u) & M_\sigma^u \\ 0 & 0 & 0 & 0 & 0 & 0 & -I \end{bmatrix}$$

and we obtain

$$B_0(y,t)Q_0 = \begin{bmatrix} A_R G\left(A_R^\top e, t\right) A_R^\top Q_C & 0 & A_V & A_E & 0 & 0 & 0 \\ -A_L^\top Q_C & 0 & 0 & 0 & 0 & 0 & 0 \\ A_V^\top Q_C & 0 & 0 & 0 & 0 & 0 & 0 \\ 0 & 0 & 0 & I & 0 & 0 & 0 \\ 0 & 0 & 0 & 0 & 0 & 0 & 0 \\ 0 & 0 & 0 & 0 & 0 & 0 & 0 \\ -\Lambda_u A_E^\top Q_C & 0 & 0 & 0 & G_u & 0 & 0 \end{bmatrix},$$

see Remark 3.33.

The next step is the calculation of $\mathcal{N}_0 \cap \mathcal{S}_0(y,t)$. This intersection is crucial for index determination and the consistent initialization as well.

Lemma 5.9. Assume Assumption 4.4, 3.31 and Property 3.32 to be satisfied. The index-1 set of the DAE (5.2) can be described by

$$\mathcal{N}_0 \cap \mathcal{S}_0\,(y,t) = \left\{ z \in \mathbb{R}^n \mid z_e \in \operatorname{im} Q_{\mathrm{CRV}},\ z_{j_V} \in \operatorname{im} Q_{\mathrm{C-V}}, \right.$$
$$\left. \Lambda_u A_E^{\mathsf{T}} Q_{\mathrm{CRV}} z_e - G_u z_{\phi_u} = z_{\widehat{\pi}_u},\ \left(z_{j_L}, z_{j_E}, z_{\widehat{a}_u} \right) = 0 \right\}.$$

Proof. For calculating the index-1 set we make use of Remark 2.27. For this we need a projector along $\operatorname{im} G_0\,(y,t)$. We are given one by

$$W_0 = \begin{bmatrix} Q_C^{\mathsf{T}} & 0 & 0 & 0 & 0 & 0 & 0 \\ 0 & 0 & 0 & 0 & 0 & 0 & 0 \\ 0 & 0 & I & 0 & 0 & 0 & 0 \\ 0 & 0 & 0 & I & 0 & 0 & 0 \\ 0 & 0 & 0 & 0 & I & 0 & 0 \\ 0 & 0 & 0 & 0 & 0 & 0 & 0 \\ 0 & 0 & 0 & 0 & 0 & 0 & 0 \end{bmatrix},$$

see Lemma C.16, and we get

$$W_0 B_0\,(y,t)\, Q_0 = \begin{bmatrix} Q_C^{\mathsf{T}} A_R G \left(A_R^{\mathsf{T}} e, t \right) A_R^{\mathsf{T}} Q_C & 0 & Q_C^{\mathsf{T}} A_V & Q_C^{\mathsf{T}} A_E & 0 & 0 & 0 \\ 0 & 0 & 0 & 0 & 0 & 0 & 0 \\ A_V^{\mathsf{T}} Q_C & 0 & 0 & 0 & 0 & 0 & 0 \\ 0 & 0 & 0 & I & 0 & 0 & 0 \\ 0 & 0 & 0 & 0 & 0 & 0 & 0 \\ 0 & 0 & 0 & 0 & 0 & 0 & 0 \\ 0 & 0 & 0 & 0 & 0 & 0 & 0 \end{bmatrix}.$$

Let be $z \in \operatorname{im} Q_0 \cap \ker W_0 B_0\,(y,t)\, Q_0$. That is true if and only if

$$z_e = Q_C z_e \tag{5.4}$$
$$z_{j_L} = 0$$
$$z_{\widehat{a}_u} = 0$$
$$z_{\widehat{\pi}_u} = \Lambda_u A_E^{\mathsf{T}} z_e - G_u z_{\phi_u}$$
$$Q_C^{\mathsf{T}} A_R G \left(A_R^{\mathsf{T}} e, t \right) A_R^{\mathsf{T}} Q_C z_e + Q_C^{\mathsf{T}} A_V z_{j_V} = 0 \tag{5.5}$$
$$A_V^{\mathsf{T}} Q_C z_e = 0 \tag{5.6}$$
$$z_{j_E} = 0$$

hold true by taking Assumption 4.4, 3.31 and Property 3.32 into account. Left-multiply of (5.5) by z_e^{T} and utilizing (5.6) leads to $Q_C z_e \in \ker A_R^{\mathsf{T}}$ due to Lemma A.3. In combination with (5.4) and (5.6) we obtain $z_e \in \operatorname{im} Q_{\mathrm{CRV}}$. Moreover, (5.5) yields $z_{j_V} \in \operatorname{im} Q_{\mathrm{C-V}}$. With it we obtain $z \in \mathcal{N}_0 \cap \mathcal{S}_0\,(y,t)$ if and only if

$$z_e \in \operatorname{im} Q_{\mathrm{CRV}}$$
$$z_{j_L} = 0$$

121

$$z_{\widehat{a}_u} = 0$$
$$z_{\widehat{\pi}_u} = \Lambda_u A_E^\top Q_{CRV} z_e - G_u z_{\phi_u}$$
$$z_{j_V} \in \text{im } Q_{C-V}$$
$$z_{j_E} = 0$$

hold true. $\qquad\qquad\qquad\qquad\qquad\qquad\qquad\qquad\qquad\qquad\qquad\qquad\qquad\qquad\qquad$ □

It is easy to see that the index-1 set $\mathcal{N}_0 \cap \mathcal{S}_0 (y, t)$ is always not empty, that is, the DAE (5.2) has never index-1. But the index does not exceed two as we will see in the next theorem.

Theorem 5.10 (index-2). Let Assumption 4.4, 5.3, 3.31 and Property 3.32 be fulfilled. The DAE (5.2) has at most index-2. It has exactly index-2 if and only if the circuit does contain a voltage source or if an EM device or if it has not a tree containing capacitors only.

Proof. For the matrix chain we need

$$G_1(y, t) = \begin{bmatrix} G(e, t) & 0 & A_V & A_E & 0 & 0 & 0 \\ -A_L^\top Q_C & L(j_L, t) & 0 & 0 & 0 & 0 & 0 \\ A_V^\top Q_C & 0 & 0 & 0 & 0 & 0 & 0 \\ 0 & 0 & 0 & I & 0 & 0 & 0 \\ 0 & 0 & 0 & 0 & 0 & 0 & 0 \\ -M_\varepsilon^u \Lambda_u A_E^\top & 0 & 0 & 0 & M_\varepsilon^u G_u & 0 & M_\varepsilon^u \\ -\Lambda_u A_E^\top Q_C & 0 & 0 & 0 & G_u & I & 0 \end{bmatrix}, \qquad (5.7)$$

with $G(e, t) = A_C C\left(A_C^\top e, t\right) A_C^\top + A_R G\left(A_R^\top e, t\right) A_R^\top Q_C$ and we calculate the index-2 set, see Definition 2.23 and Remark 2.28. For the subspaces needed, we have to provide a projector along $\text{im } G_1(y, t)$. We are given one by

$$W_1 = \begin{bmatrix} Q_{CRV}^\top & 0 & 0 & -Q_{CRV}^\top A_E & 0 & 0 & 0 \\ 0 & 0 & 0 & 0 & 0 & 0 & 0 \\ 0 & 0 & Q_{C-V}^\top & 0 & 0 & 0 & 0 \\ 0 & 0 & 0 & 0 & 0 & 0 & 0 \\ 0 & 0 & 0 & 0 & I & 0 & 0 \\ 0 & 0 & 0 & 0 & 0 & 0 & 0 \\ 0 & 0 & 0 & 0 & 0 & 0 & 0 \end{bmatrix},$$

see Lemma C.17. Next we take into account

$$P_0 = \begin{bmatrix} P_C & 0 & 0 & 0 & 0 & 0 & 0 \\ 0 & I & 0 & 0 & 0 & 0 & 0 \\ 0 & 0 & 0 & 0 & 0 & 0 & 0 \\ 0 & 0 & 0 & 0 & 0 & 0 & 0 \\ 0 & 0 & 0 & 0 & 0 & 0 & 0 \\ 0 & 0 & 0 & 0 & 0 & I & 0 \\ -\Lambda_u A_E^\top Q_C & 0 & 0 & 0 & G_u & 0 & I \end{bmatrix},$$

where P_0 is the complementary projector to Q_0, and we calculate

$$B_0(y,t)P_0 = \begin{bmatrix} A_R G\left(A_R^\top e, t\right) A_R^\top P_C & A_L & 0 & 0 & 0 & 0 & 0 \\ -A_L^\top P_C & 0 & 0 & 0 & 0 & 0 & 0 \\ A_V^\top P_C & 0 & 0 & 0 & 0 & 0 & 0 \\ 0 & 0 & 0 & 0 & 0 & -\Lambda_u^\top K_{\nu,d}^u(\widehat{a}_u) & 0 \\ 0 & 0 & 0 & 0 & 0 & \widetilde{S}_u M_{\overline{\nu}}^u & 0 \\ -M_\sigma^u \Lambda_u A_E^\top & 0 & 0 & 0 & M_\sigma^u G_u & K_{\nu,d}^u(\widehat{a}_u) & M_\sigma^u \\ \Lambda_u A_E^\top Q_C & 0 & 0 & 0 & -G_u & 0 & -I \end{bmatrix}$$

and

$$W_1 B_0(y,t)P_0 = \begin{bmatrix} 0 & Q_{CRV}^\top A_L & 0 & 0 & 0 & Q_{CRV}^\top A_E \Lambda_u^\top K_{\nu,d}^u(\widehat{a}_u) & 0 \\ 0 & 0 & 0 & 0 & 0 & 0 & 0 \\ Q_{C-V}^\top A_V^\top P_C & 0 & 0 & 0 & 0 & 0 & 0 \\ 0 & 0 & 0 & 0 & 0 & 0 & 0 \\ 0 & 0 & 0 & 0 & 0 & \widetilde{S}_u M_{\overline{\nu}}^u & 0 \\ 0 & 0 & 0 & 0 & 0 & 0 & 0 \\ 0 & 0 & 0 & 0 & 0 & 0 & 0 \end{bmatrix}.$$

Let be $z \in \ker G_1(y,t) \cap \ker W_1 B_0(y,t)P_0$. That is true if and only if the conditions

$$Q_C z_e = Q_{CRV} z_e \tag{5.8}$$

$$z_{jv} = Q_{C-V} z_{jv} \tag{5.9}$$

$$z_{jE} = 0$$

$$L(j_L,t)^{-1} A_L^\top Q_C z_e = z_{j_L} \tag{5.10}$$

$$P_C z_e = -H_C\left(A_C^\top e, t\right)^{-1} A_V Q_{C-V} z_{jv} \tag{5.11}$$

$$\Lambda_u A_E^\top z_e - G_u z_{\phi_u} = z_{\widehat{\pi}_u} \tag{5.12}$$

$$\Lambda_u A_E^\top Q_C z_e - G_u z_{\phi_u} = z_{\widehat{a}_u} \tag{5.13}$$

$$Q_{CRV}^\top A_L z_{j_L} + Q_{CRV}^\top A_E \Lambda_u^\top K_{\nu,d}^u(\widehat{a}_u) z_{\widehat{a}_u} = 0 \tag{5.14}$$

$$Q_{C-V}^\top A_V^\top P_C z_e = 0 \tag{5.15}$$

$$\widetilde{S}_u M_{\overline{\nu}}^u z_{\widehat{a}_u} = 0 \tag{5.16}$$

are fulfilled, using the representation

$$\mathcal{N}_1(y,t) = \left\{ z \in \mathbb{R}^n \mid Q_C z_e \in \operatorname{im} Q_{CRV}, \; z_{jv} \in \operatorname{im} Q_{C-V}, \right.$$
$$L(j_L,t)^{-1} A_L^\top Q_C z_e = z_{j_L}, \; P_C z_e = -H_C\left(A_C^\top e, t\right)^{-1} A_V Q_{C-V} z_{jv}, \; z_{jE} = 0,$$
$$\left. \Lambda_u A_E^\top z_e - G_u z_{\phi_u} = z_{\widehat{\pi}_u}, \; \Lambda_u A_E^\top Q_C z_e - G_u z_{\phi_u} = z_{\widehat{a}_u} \right\},$$

see Lemma C.18, where $H_C\left(A_C^\top e, t\right) = A_C C\left(A_C^\top e, t\right) A_C^\top + Q_C^\top Q_C$ is positive definite, see Lemma A.10, using Assumption 4.4, 3.31 and Property 3.32. Left-multiplying of (5.11) by $z_{jv}^\top Q_{C-V}^\top A_V^\top$ and using (5.15) leads to

$$z_{jv}^\top Q_{C-V}^\top A_V^\top H_C\left(A_C^\top e, t\right)^{-1} A_V Q_{C-V} z_{jv} = 0.$$

With Lemma A.3 and (5.9) we deduce

$$A_V z_{j_V} = 0 \text{ and } z_{j_V} = 0$$

since V-loops are forbidden. From (5.11) we acquire $z_e \in \operatorname{im} Q_C$ and together with (5.8) we get $z_e \in \operatorname{im} Q_{CRV}$. Combining (5.10), (5.13) and (5.14) yields

$$Q_{CRV}^\top A_L L \left(j_L, t\right)^{-1} A_L^\top Q_{CRV} z_e + Q_{CRV}^\top A_E \Lambda_u^\top K_{\nu,d}^u (\widehat{a}_u) \Lambda_u A_E^\top Q_{CRV} z_e = 0,$$

since $C_u G_u = 0$ holds true, and we deduce $z_e \in \ker \begin{bmatrix} A_L & A_E \end{bmatrix}^\top$. Thus we come by the condition

$$z_e \in \ker \begin{bmatrix} A_C & A_R & A_L & A_E & A_V \end{bmatrix}^\top.$$

Because I-cutsets are forbidden, we gain $z_e = 0$. Then the relation (5.10) leads to $z_{j_L} = 0$. Left-multiplying (5.13) by $\widetilde{S}_u M_{\bar{\nu}}^u$ and using (5.16) yields

$$\widetilde{S}_u M_{\bar{\nu}}^u G_u z_{\phi_u} = 0$$

and hence $z_{\phi_u} = 0$ due to the choice of $M_{\bar{\nu}}^u = M_\zeta^u G_u M_\xi^u \widetilde{S}_u M_\zeta^u$. From (5.12) and (5.13) we get $\left(z_{\widehat{a}_u}, z_{\widehat{\pi}_u}\right) = 0$ and we conclude $z = 0$, see Definition 2.23. □

To start the integration of the DAE (5.2) we need a consistent initialization. For the index-2 case we apply Theorem 2.58. For this we need to check the requirements.

Assumption 5.11. For the DAE (5.2) exist the continuous partial derivatives $\frac{\partial}{\partial t} d\left(y, t\right)$ and $\frac{\partial}{\partial t} W_1 b\left(y, t\right)$ for all $(y, t) \in \mathcal{D} \times \mathcal{I}$.

These assumptions are not a restriction since, if a solution exists, then $\frac{\partial}{\partial t} d\left(y, t\right)$ exists and is continuous. Moreover $W_1 b\left(y, t\right)$ describes exactly the hidden constraints and hence $\frac{\partial}{\partial t} W_1 b\left(y, t\right)$ needs to exists and to be continuous to have a solution of the problem. In addition, the DAE (5.2) has a constant matrix A and there are the constant projectors Q_0 and W_1. It remains to show that the index-2 variables enter linearly only.

Lemma 5.12. Let Assumption 4.4, 3.31 and Property 3.32 be fulfilled. The index-2 variables enter the DAE (5.2) linearly only.

Proof. From Lemma 5.9 we easily obtain a constant projector T onto $\mathcal{N}_0 \cap \mathcal{S}_0\left(y, t\right)$ given by

$$T = \begin{bmatrix} Q_{CRV} & 0 & 0 & 0 & 0 & 0 & 0 \\ 0 & 0 & 0 & 0 & 0 & 0 & 0 \\ 0 & 0 & Q_{C-V} & 0 & 0 & 0 & 0 \\ 0 & 0 & 0 & 0 & 0 & 0 & 0 \\ 0 & 0 & 0 & 0 & I & 0 & 0 \\ 0 & 0 & 0 & 0 & 0 & 0 & 0 \\ \Lambda_u A_E^\top Q_{CRV} & 0 & 0 & 0 & -G_u & 0 & 0 \end{bmatrix}.$$

Furthermore the complementary projector U is given by

$$U = \begin{bmatrix} P_{CRV} & 0 & 0 & 0 & 0 & 0 & 0 \\ 0 & I & 0 & 0 & 0 & 0 & 0 \\ 0 & 0 & P_{C-V} & 0 & 0 & 0 & 0 \\ 0 & 0 & 0 & I & 0 & 0 & 0 \\ 0 & 0 & 0 & 0 & 0 & 0 & 0 \\ 0 & 0 & 0 & 0 & 0 & I & 0 \\ -\Lambda_u A_E^{\mathsf{T}} Q_{CRV} & 0 & 0 & 0 & G_u & 0 & I \end{bmatrix}$$

and the unknowns are divided into

$$y = Ty + Uy = \begin{pmatrix} Q_{CRV}e \\ 0 \\ Q_{C-v}j_v \\ 0 \\ \phi_u \\ 0 \\ \Lambda_u A_E^{\mathsf{T}} Q_{CRV}e - G_u\phi_u \end{pmatrix} + \begin{pmatrix} P_{CRV}e \\ j_L \\ P_{C-v}j_v \\ j_E \\ 0 \\ \widehat{a}_u \\ -\Lambda_u A_E^{\mathsf{T}} Q_{CRV}e + G_u\phi_u + \widehat{\pi}_u \end{pmatrix}.$$

Now we can write $b(y,t) = b(Uy,t) + BTy$ with

$$B = \begin{bmatrix} 0 & 0 & A_V & 0 & 0 & 0 & 0 \\ -A_L^{\mathsf{T}} & 0 & 0 & 0 & 0 & 0 & 0 \\ 0 & 0 & 0 & 0 & 0 & 0 & 0 \\ 0 & 0 & 0 & 0 & 0 & 0 & 0 \\ 0 & 0 & 0 & 0 & 0 & 0 & 0 \\ 0 & 0 & 0 & 0 & 0 & 0 & 0 \\ 0 & 0 & 0 & 0 & 0 & 0 & -I \end{bmatrix},$$

using Lemma 4.21 without memristors. The relation $d(y,t) = d(Uy,t)$ is obvious by Lemma 2.54. $\qquad\square$

The DAE (5.2) fulfills all requirements to apply Theorem 2.58 in case of index-2. But we still need an operating point when we want to integrate it numerically. Since Theorem 2.58 is applicable to the DAE an operating point is sufficient to start the numerical integration, see Lemma 2.61.

Let be $z^0 = \left(z_C^0, z_L^0, z_{\widehat{a}_u}^0, z_{\widehat{\pi}_u}^0\right)$, $y^0 = (e^0, j_L^0, j_V^0, j_E^0, \phi_u^0, \widehat{a}_u^0, \widehat{\pi}_u^0)$ and $t_0 \in \mathcal{I}$. We choose $\widehat{a}_u^0 \in \ker C_u$ and (ϕ_u^0, π_u^0) arbitrarily. Then we get

$$j_E^0 = 0$$

and choose z_C^0 and j_L^0 such that

$$-A_C z_C^0 - A_L j_L^0 - A_I i_s(t_0) \in \operatorname{im} \begin{bmatrix} A_R & A_V \end{bmatrix},$$

see Remark 4.24 without memristors. Next we determine (e^0, j_V^0). For that we need a solution of the nonlinear system:

$$A_{RGR}\left(A_R^\top e^0, t_0\right) + A_V j_V^0 = -A_C z_C^0 - A_L j_L^0 - A_I i_s\left(t_0\right)$$
$$A_V^\top e^0 = v_s\left(t_0\right) \tag{5.17}$$

To obtain a solution (e^0, j_V^0) of (5.17) we apply Approach 4.25 without memristors. Then we compute the missing parts by:

$$z_L^0 = A_L^\top e^0$$
$$M_\varepsilon^u z_{\widehat{a}_u}^0 = M_\sigma^u \Lambda_u A_E^\top e^0 - M_\sigma^u G_u \phi_u^0 - M_\sigma^u \widehat{\pi}_u^0$$
$$z_{\widehat{\pi}_u}^0 = \widehat{\pi}_u^0$$

Remark 5.13. Due to the structure of the DAE (5.2) we obtain a locally unique solution through every consistent initial value and the perturbation index-2, see Theorem 2.46 and 2.50.

5.3.2 Field/Circuit System using Lorenz Gauge

The coupled system (5.1) using Lorenz gauge, that is, $\vartheta = 1$, can be formulated as a DAE given by

$$A\frac{d}{dt}d(y, t) + b(y, t) = 0 \tag{5.18}$$

with unknowns $y = (e, j_L, j_V, j_E, \phi_u, \widehat{a}_u, \widehat{\pi}_u)$ and the describing matrix and functions

$$A = \begin{bmatrix} A_C & 0 & 0 & 0 & 0 \\ 0 & I & 0 & 0 & 0 \\ 0 & 0 & 0 & 0 & 0 \\ 0 & 0 & 0 & 0 & 0 \\ 0 & 0 & \widetilde{S}_u M_\varepsilon^u G_u & 0 & 0 \\ 0 & 0 & M_\varepsilon^u G_u & 0 & M_\varepsilon^u \\ 0 & 0 & 0 & I & 0 \end{bmatrix}, \quad d(y, t) = \begin{pmatrix} q_C\left(A_C^\top e, t\right) \\ \phi_L\left(j_L, t\right) \\ \phi_u \\ \widehat{a}_u \\ -\Lambda_u A_E^\top e + \widehat{\pi}_u \end{pmatrix}$$

as well as

$$b(y, t) = \begin{pmatrix} A_{RGR}\left(A_R^\top e, t\right) + A_L j_L + A_V j_V + A_E j_E + A_I i_s\left(t\right) \\ -A_L^\top e \\ A_V^\top e - v_s\left(t\right) \\ j_E - \Lambda_u^\top K_\nu^u(\widehat{a}_u)\widehat{a}_u \\ \widetilde{S}_u M_\nu^u \widehat{a}_u \\ -M_\sigma^u \Lambda_u A_E^\top e + M_\sigma^u G_u \phi_u + K_\nu^u(\widehat{a}_u)\widehat{a}_u + M_\sigma^u \widehat{\pi}_u \\ -\widehat{\pi}_u \end{pmatrix}.$$

Lemma 5.14. Let Assumption 4.4 and 3.31 be satisfied. Then, the DAE (5.18) has a properly stated leading term where the constant projector

$$R = \begin{bmatrix} A_C^+ A_C & 0 \\ 0 & I \end{bmatrix}$$

realizes the decomposition (2.9).

Proof. We follow the idea of the proof of Lemma 5.7. $\quad\square$

The first step is an index-0 result. For this we need the matrix, see Definition 2.21,

$$G_0(y,t) = \begin{bmatrix} A_C C\left(A_C^T e,t\right) A_C^T & 0 & 0 & 0 & 0 & 0 & 0 \\ 0 & L(j_L,t) & 0 & 0 & 0 & 0 & 0 \\ 0 & 0 & 0 & 0 & 0 & 0 & 0 \\ 0 & 0 & 0 & 0 & 0 & 0 & 0 \\ 0 & 0 & 0 & 0 & \tilde{S}_u M_\varepsilon^u G_u & 0 & 0 \\ -M_\varepsilon^u \Lambda_u A_E^T & 0 & 0 & 0 & M_\varepsilon^u G_u & 0 & M_\varepsilon^u \\ 0 & 0 & 0 & 0 & 0 & I & 0 \end{bmatrix} \quad (5.19)$$

with

$$D(y,t) = \begin{bmatrix} C\left(A_C^T e,t\right) A_C^T & 0 & 0 & 0 & 0 & 0 & 0 \\ 0 & L(j_L,t) & 0 & 0 & 0 & 0 & 0 \\ 0 & 0 & 0 & 0 & I & 0 & 0 \\ 0 & 0 & 0 & 0 & 0 & I & 0 \\ -\Lambda_u A_E^T & 0 & 0 & 0 & 0 & 0 & I \end{bmatrix}.$$

Theorem 5.15 (index-0). Let Assumption 4.4, 3.31 and Property 3.32 be fulfilled. The DAE (5.18) has index-0 if and only if the circuit does not contain voltage sources and if EM device and if there is a tree containing capacitors only.

Proof. Following the proof of Theorem 4.13, the remaining zero rows and columns disappear if and only if there is no EM device. $\quad\square$

The next step is to describe the intersection index-1 set. For this we compute a projector onto $\ker G_0(y,t)$ and the derivative

$$B_0(y,t) = \begin{bmatrix} A_R G\left(A_R^T e,t\right) A_R^T & A_L & A_V & A_E & 0 & 0 & 0 \\ -A_L^T & 0 & 0 & 0 & 0 & 0 & 0 \\ A_V^T & 0 & 0 & 0 & 0 & 0 & 0 \\ 0 & 0 & 0 & I & 0 & -\Lambda_u^T K_{\nu,d}^u(\hat{a}_u) & 0 \\ 0 & 0 & 0 & 0 & 0 & \tilde{S}_u M_{\bar\nu}^u & 0 \\ -M_\sigma^u \Lambda_u A_E^T & 0 & 0 & 0 & M_\sigma^u G_u & K_{\nu,d}^u(\hat{a}_u) & M_\sigma^u \\ 0 & 0 & 0 & 0 & 0 & 0 & -I \end{bmatrix}.$$

Let be $z \in \ker G_0(y,t)$. That is true if and only if

$$z_e \in \operatorname{im} Q_C$$

127

$$z_{j_L} = 0$$
$$z_{\widehat{\pi}_u} = \Lambda_u A_E^\top Q_C z_e$$
$$z_{\phi_u} = 0$$
$$z_{\widehat{a}_u} = 0$$

due to Lemma A.3. Hence we can choose a projector onto $\ker G_0(y, t)$ by

$$Q_0 = \begin{bmatrix} Q_C & 0 & 0 & 0 & 0 & 0 & 0 \\ 0 & 0 & 0 & 0 & 0 & 0 & 0 \\ 0 & 0 & I & 0 & 0 & 0 & 0 \\ 0 & 0 & 0 & I & 0 & 0 & 0 \\ 0 & 0 & 0 & 0 & 0 & 0 & 0 \\ 0 & 0 & 0 & 0 & 0 & 0 & 0 \\ \Lambda_u A_E^\top Q_C & 0 & 0 & 0 & 0 & 0 & 0 \end{bmatrix}$$

and we calculate

$$B_0(y, t) Q_0 = \begin{bmatrix} A_R G\left(A_R^\top e, t\right) A_R^\top Q_C & 0 & A_V & A_E & 0 & 0 & 0 \\ -A_L^\top Q_C & 0 & 0 & 0 & 0 & 0 & 0 \\ A_V^\top Q_C & 0 & 0 & 0 & 0 & 0 & 0 \\ 0 & 0 & 0 & I & 0 & 0 & 0 \\ 0 & 0 & 0 & 0 & 0 & 0 & 0 \\ 0 & 0 & 0 & 0 & 0 & 0 & 0 \\ -\Lambda_u A_E^\top Q_C & 0 & 0 & 0 & 0 & 0 & 0 \end{bmatrix},$$

see Remark 3.33. Next $\mathcal{N}_0 \cap \mathcal{S}_0(y, t)$ is calculated, since the intersection plays an important role for the index calculation and for the consistent initialization.

Lemma 5.16. Let Assumption 4.4, 3.31 and Property 3.32 hold true. The index-1 set of the DAE (5.18) can be described by

$$\mathcal{N}_0 \cap \mathcal{S}_0(y, t) = \big\{ z \in \mathbb{R}^n \mid z_e \in \operatorname{im} Q_{CRV}, \ z_{j_V} \in \operatorname{im} Q_{C-V},$$
$$\Lambda_u A_E^\top Q_{CRV} z_e = z_{\widehat{\pi}_u}, \ \big(z_{j_L}, z_{j_E}, z_{\phi_u}, z_{\widehat{a}_u}\big) = 0 \big\}.$$

Proof. For calculating the index-1 set we make use of Remark 2.27. For that we need a projector along $\operatorname{im} G_0(y, t)$. We are given one by

$$W_0 = \begin{bmatrix} Q_C^\top & 0 & 0 & 0 & 0 & 0 & 0 \\ 0 & 0 & 0 & 0 & 0 & 0 & 0 \\ 0 & 0 & I & 0 & 0 & 0 & 0 \\ 0 & 0 & 0 & I & 0 & 0 & 0 \\ 0 & 0 & 0 & 0 & 0 & 0 & 0 \\ 0 & 0 & 0 & 0 & 0 & 0 & 0 \\ 0 & 0 & 0 & 0 & 0 & 0 & 0 \end{bmatrix},$$

see Lemma C.19, and we get

$$
W_0 B_0\left(y,t\right) Q_0 =
\begin{bmatrix}
Q_C^T A_R G\left(A_R^T e,t\right) A_R^T Q_C & 0 & Q_C^T A_V & Q_C^T A_E & 0 & 0 & 0 \\
0 & 0 & 0 & 0 & 0 & 0 & 0 \\
A_V^T Q_C & 0 & 0 & 0 & 0 & 0 & 0 \\
0 & 0 & 0 & I & 0 & 0 & 0 \\
0 & 0 & 0 & 0 & 0 & 0 & 0 \\
0 & 0 & 0 & 0 & 0 & 0 & 0 \\
0 & 0 & 0 & 0 & 0 & 0 & 0
\end{bmatrix}.
$$

The rest of the proof is entirely analog to the proof of Lemma 5.9. $\qquad\square$

With the characterization of the intersection we are able to deduce network topological index-1 conditions for the coupled system DAE (5.18). The EM devices are insert into the circuit as a kind of controlled current sources, but the analysis show that for using Lorenz gauge they, from the index point of view, behave like inductances.

Theorem 5.17 (index-1). Let Assumption 4.4, 3.31 and Property 3.32 be true. The DAE (5.18) has index-1 if and only if there is at least a voltage sources in the circuit or there is no tree containing capacitors only and if there is neither an LEI-cutset nor a CV-loop with at least one voltage source.

Proof. We make use of the representation of $\mathcal{N}_0 \cap \mathcal{S}_0\left(y,t\right)$ as proposed in Lemma 5.16. The intersection $\mathcal{N}_0 \cap \mathcal{S}_0\left(y,t\right)$ is trivial if and only if $Q_{CRV} = 0$ and $Q_{C-V} = 0$. This is equivalent to the circuit containing neither LEI-cutsets nor CV-loops, see Lemma 5.5 and 5.6. Using Definition 2.23 we get the DAE (5.18) to be of index-1. $\qquad\square$

The DAE (5.18) can be of index-2 also, but higher index problems can be avoided. We will see that LEI-cutsets and CV-loops are the only critical circuit configurations. Obviously the dimension of $\mathcal{N}_0 \cap \mathcal{S}_0\left(x,t\right)$ is constant, which is important for the index-2 case.

Theorem 5.18 (index-2). Let Assumption 4.4, 5.3, 3.31 and Property 3.32 be fulfilled. The DAE (5.18) has index-2 if and only if there is an LEI-cutset or a CV-loop with at least one voltage source.

Proof. For the matrix chain we need

$$
G_1\left(y,t\right) =
\begin{bmatrix}
G\left(e,t\right) & 0 & A_V & A_E & 0 & 0 & 0 \\
-A_L^T Q_C & L\left(j_L,t\right) & 0 & 0 & 0 & 0 & 0 \\
A_V^T Q_C & 0 & 0 & 0 & 0 & 0 & 0 \\
0 & 0 & 0 & I & 0 & 0 & 0 \\
0 & 0 & 0 & 0 & \widetilde{S}_u M_\varepsilon^u G_u & 0 & 0 \\
-M_\varepsilon^u \Lambda_u A_E^T & 0 & 0 & 0 & M_\varepsilon^u G_u & 0 & M_\varepsilon^u \\
-\Lambda_u A_E^T Q_C & 0 & 0 & 0 & 0 & I & 0
\end{bmatrix},
\tag{5.20}
$$

with $G(e, t) = A_C C\left(A_C^\mathsf{T} e, t\right) A_C^\mathsf{T} + A_R G\left(A_R^\mathsf{T} e, t\right) A_R^\mathsf{T} Q_C$, and we calculate the index-2 set, see Definition 2.23 and Remark 2.28. For the subspaces needed, we have to provide a projector along $\operatorname{im} G_1(y, t)$. We are given one by

$$
W_1 = \begin{bmatrix}
Q_{CRV}^\mathsf{T} & 0 & 0 & -Q_{CRV}^\mathsf{T} A_E & 0 & 0 & 0 \\
0 & 0 & 0 & 0 & 0 & 0 & 0 \\
0 & 0 & Q_{C-V}^\mathsf{T} & 0 & 0 & 0 & 0 \\
0 & 0 & 0 & 0 & 0 & 0 & 0 \\
0 & 0 & 0 & 0 & 0 & 0 & 0 \\
0 & 0 & 0 & 0 & 0 & 0 & 0 \\
0 & 0 & 0 & 0 & 0 & 0 & 0
\end{bmatrix},
$$

see Lemma C.20. Next we take into account

$$
P_0 = \begin{bmatrix}
P_C & 0 & 0 & 0 & 0 & 0 & 0 \\
0 & I & 0 & 0 & 0 & 0 & 0 \\
0 & 0 & 0 & 0 & 0 & 0 & 0 \\
0 & 0 & 0 & 0 & 0 & 0 & 0 \\
0 & 0 & 0 & 0 & I & 0 & 0 \\
0 & 0 & 0 & 0 & 0 & I & 0 \\
-\Lambda_u A_E^\mathsf{T} Q_C & 0 & 0 & 0 & 0 & 0 & I
\end{bmatrix},
$$

where P_0 is the complementary projector to Q_0, and we calculate

$$
B_0(y, t) P_0 = \begin{bmatrix}
A_R G\left(A_R^\mathsf{T} e, t\right) A_R^\mathsf{T} P_C & A_L & 0 & 0 & 0 & 0 & 0 \\
-A_L^\mathsf{T} P_C & 0 & 0 & 0 & 0 & 0 & 0 \\
A_V^\mathsf{T} P_C & 0 & 0 & 0 & 0 & 0 & 0 \\
0 & 0 & 0 & 0 & 0 & -\Lambda_u^\mathsf{T} K_{\nu,d}^u(\widehat{a}_u) & 0 \\
0 & 0 & 0 & 0 & 0 & \widetilde{S}_u M_{\bar{\nu}}^u & 0 \\
-M_\sigma^u \Lambda_u A_E^\mathsf{T} & 0 & 0 & 0 & M_\sigma^u G_u & K_{\nu,d}^u(\widehat{a}_u) & M_\sigma^u \\
\Lambda_u A_E^\mathsf{T} Q_C & 0 & 0 & 0 & -G_u & 0 & -I
\end{bmatrix}
$$

and

$$
W_1 B_0(y, t) P_0 = \begin{bmatrix}
0 & Q_{CRV}^\mathsf{T} A_L & 0 & 0 & 0 & Q_{CRV}^\mathsf{T} \Lambda_u^\mathsf{T} K_{\nu,d}^u(\widehat{a}_u) & 0 \\
0 & 0 & 0 & 0 & 0 & 0 & 0 \\
Q_{C-V}^\mathsf{T} A_V^\mathsf{T} P_C & 0 & 0 & 0 & 0 & 0 & 0 \\
0 & 0 & 0 & 0 & 0 & 0 & 0 \\
0 & 0 & 0 & 0 & 0 & 0 & 0 \\
0 & 0 & 0 & 0 & 0 & 0 & 0 \\
0 & 0 & 0 & 0 & 0 & 0 & 0
\end{bmatrix}.
$$

Let be $z \in \ker G_1(y, t) \cap \ker W_1 B_0(y, t) P_0$. That is true if and only if the conditions

$$
Q_C z_e \in \operatorname{im} Q_{CRV} \tag{5.21}
$$

$$
z_{jv} \in \operatorname{im} Q_{C-V} \tag{5.22}
$$

$$z_{j_E} = 0$$
$$z_{\phi_u} = 0$$
$$L\left(j_L, t\right)^{-1} A_L^{\mathsf{T}} Q_C z_e = z_{j_L} \tag{5.23}$$
$$P_C z_e = -H_C \left(A_C^{\mathsf{T}} e, t\right)^{-1} A_V Q_{C-V} z_{j_V} \tag{5.24}$$
$$\Lambda_u A_E^{\mathsf{T}} z_e = z_{\widehat{\pi}_u}$$
$$\Lambda_u A_E^{\mathsf{T}} Q_C z_e = z_{\widehat{a}_u} \tag{5.25}$$
$$Q_{CRV}^{\mathsf{T}} A_L z_{j_L} + Q_{CRV}^{\mathsf{T}} \Lambda_u^{\mathsf{T}} K_{\nu,d}^u(\widehat{a}_u) z_{\widehat{a}_u} = 0 \tag{5.26}$$
$$Q_{C-V}^{\mathsf{T}} A_V^{\mathsf{T}} P_C z_e = 0 \tag{5.27}$$

are fulfilled, using

$$\mathcal{N}_1\left(y, t\right) = \big\{ z \in \mathbb{R}^n \mid Q_C z_e \in \operatorname{im} Q_{CRV},\ z_{j_V} \in \operatorname{im} Q_{C-V},$$
$$L\left(j_L, t\right)^{-1} A_L^{\mathsf{T}} Q_C z_e = z_{j_L},\ P_C z_e = -H_C \left(A_C^{\mathsf{T}} e, t\right)^{-1} A_V Q_{C-V} z_{j_V},$$
$$\left(z_{j_E}, z_{\phi_u}\right) = 0,\ \Lambda_u A_E^{\mathsf{T}} z_e = z_{\widehat{\pi}_u},\ \Lambda_u A_E^{\mathsf{T}} Q_C z_e = z_{\widehat{a}_u} \big\},$$

see Lemma C.21, where $H_C\left(A_C^{\mathsf{T}} e, t\right) = A_C C\left(A_C^{\mathsf{T}} e, t\right) A_C^{\mathsf{T}} + Q_C^{\mathsf{T}} Q_C$ is positive definite, see Lemma A.10, using Assumption 4.4, 3.31 and Property 3.32. Left-multiplying of (5.24) by $z_{j_V}^{\mathsf{T}} Q_{C-V}^{\mathsf{T}} A_V^{\mathsf{T}}$ and using (5.27) leads to

$$z_{j_V}^{\mathsf{T}} Q_{C-V}^{\mathsf{T}} A_V^{\mathsf{T}} H_C \left(A_C^{\mathsf{T}} e, t\right)^{-1} A_V Q_{C-V} z_{j_V} = 0.$$

With Lemma A.3 and (5.22) we deduce

$$A_V z_{j_V} = 0 \text{ and } z_{j_V} = 0$$

since V-loops are forbidden. From (5.24) we acquire $z_e \in \operatorname{im} Q_C$ and together with (5.21) we come by $z_e \in \operatorname{im} Q_{CRV}$. Combining (5.23), (5.25) and (5.26) yields

$$Q_{CRV}^{\mathsf{T}} A_L L\left(j_L, t\right)^{-1} A_L^{\mathsf{T}} Q_{CRV} z_e + Q_{CRV}^{\mathsf{T}} A_E \Lambda_u^{\mathsf{T}} K_{\nu,d}^u(\widehat{a}_u) \Lambda_u A_E^{\mathsf{T}} Q_{CRV} z_e = 0$$

since $C_u G_u = 0$ holds true, and we deduce $z_e \in \ker\begin{bmatrix} A_L & A_E \end{bmatrix}^{\mathsf{T}}$. Thus we come by the condition

$$z_e \in \ker\begin{bmatrix} A_C & A_R & A_L & A_E & A_V \end{bmatrix}^{\mathsf{T}}.$$

Because I-cutsets are forbidden we gain $z_e = 0$. Then the relations (5.23), (5.12) and (5.25) leads $\left(z_{j_L}, z_{\widehat{a}_u}, z_{\widehat{\pi}_u}\right) = 0$ and we conclude $z = 0$, see Definition 2.23. □

The topological index results for the coupled system using Lorenz gauge are also a extentsion of the topological index results of [BBS11] for coupled MQS device/circuit systems using Lorenz gauge.

In order to start the integration of the DAE (5.18) we need a consistent initialization. In case of index-1 we make direct use of Theorem 2.51. In case of index-2 we apply Theorem 2.58. For this we need to check the requirements.

Assumption 5.19. For the DAE (5.18) exist the continuous partial derivatives $\frac{\partial}{\partial t}d\,(y,t)$ and $\frac{\partial}{\partial t}W_1b\,(y,t)$ for all $(y,t) \in \mathcal{D} \times \mathcal{I}$.

These assumptions are not a restriction since, if a solution exists, then $\frac{\partial}{\partial t}d\,(y,t)$ exists and is continuous. Moreover $W_1b\,(y,t)$ describes exactly the hidden constraints and hence $\frac{\partial}{\partial t}W_1b\,(y,t)$ needs to exists and to be continuous in order to have a solution of the problem.

In addition, the DAE (5.18) has a constant matrix A and there are constant projectors Q_0 and W_1. It remains to show that the index-2 variables enter linearly only.

Lemma 5.20. Let Assumption 4.4, 3.31 and Property 3.32 be fulfilled. The index-2 variables enter the DAE (5.18) linearly only.

Proof. From Lemma 5.16 we easily obtain a constant projector T onto $\mathcal{N}_0 \cap \mathcal{S}_0\,(y,t)$ given by

$$
T = \begin{bmatrix}
Q_{\mathrm{CRV}} & 0 & 0 & 0 & 0 & 0 & 0 \\
0 & 0 & 0 & 0 & 0 & 0 & 0 \\
0 & 0 & Q_{\mathrm{C-V}} & 0 & 0 & 0 & 0 \\
0 & 0 & 0 & 0 & 0 & 0 & 0 \\
0 & 0 & 0 & 0 & 0 & 0 & 0 \\
0 & 0 & 0 & 0 & 0 & 0 & 0 \\
\Lambda_u A_E^\top Q_{\mathrm{CRV}} & 0 & 0 & 0 & 0 & 0 & 0
\end{bmatrix}.
$$

Furthermore, the complementary projector U is given by

$$
U = \begin{bmatrix}
P_{\mathrm{CRV}} & 0 & 0 & 0 & 0 & 0 & 0 \\
0 & I & 0 & 0 & 0 & 0 & 0 \\
0 & 0 & P_{\mathrm{C-V}} & 0 & 0 & 0 & 0 \\
0 & 0 & 0 & I & 0 & 0 & 0 \\
0 & 0 & 0 & 0 & I & 0 & 0 \\
0 & 0 & 0 & 0 & 0 & I & 0 \\
-\Lambda_u A_E^\top Q_{\mathrm{CRV}} & 0 & 0 & 0 & 0 & 0 & I
\end{bmatrix}
$$

and the unknowns are divided into

$$
y = Ty + Uy = \begin{pmatrix}
Q_{\mathrm{CRV}}e \\
0 \\
Q_{\mathrm{C-v}}j_v \\
0 \\
0 \\
0 \\
\Lambda_u A_E^\top Q_{\mathrm{CRV}}e
\end{pmatrix} + \begin{pmatrix}
P_{\mathrm{CRV}}e \\
j_L \\
P_{\mathrm{C-v}}j_v \\
j_E \\
\phi_u \\
\widehat{a}_u \\
-\Lambda_u A_E^\top Q_{\mathrm{CRV}}e + \widehat{\pi}_u
\end{pmatrix}.
$$

Now we can write $b(y,t) = b(Uy,t) + BTy$ with

$$
B = \begin{bmatrix}
0 & 0 & A_V & 0 & 0 & 0 & 0 \\
-A_L^\top & 0 & 0 & 0 & 0 & 0 & 0 \\
0 & 0 & 0 & 0 & 0 & 0 & 0 \\
0 & 0 & 0 & 0 & 0 & 0 & 0 \\
0 & 0 & 0 & 0 & 0 & 0 & 0 \\
0 & 0 & 0 & 0 & 0 & 0 & 0 \\
0 & 0 & 0 & 0 & 0 & 0 & -I
\end{bmatrix},
$$

with Lemma 4.21 without memristors. The relation $d(y,t) = d(Uy,t)$ is obvious by Lemma 2.54. □

The DAE (5.18) fulfills all requirements to apply Theorem 2.58 in case of index-2. But we still need an operating point when we want to integrate it numerically. Since Theorem 2.58 is applicable to the DAE an operating point is sufficient to start the numerical integration, see Lemma 2.61.

Let be $z^0 = \left(z_C^0, z_L^0, z_{\phi_u}^0, z_{\widehat{a}_u}^0, z_{\widehat{\pi}_u}^0\right)$, $y^0 = (e^0, j_L^0, j_V^0, j_E^0, \phi_u^0, \widehat{a}_u^0, \widehat{\pi}_u^0)$ and $t_0 \in \mathcal{I}$. We choose $\widehat{a}_u^0 \in \ker C_u$ and $(\phi_u^0, \widehat{\pi}_u^0)$ arbitrarily. Then we get

$$
j_E^0 = 0
$$

and choose z_C^0 and j_L^0 such that

$$
-A_C z_C^0 - A_L j_L^0 - A_I i_s(t_0) \in \text{im}\begin{bmatrix} A_R & A_V \end{bmatrix},
$$

see Remark 4.24 without memristors. Next we determine (e^0, j_V^0). For that we need a solution of the nonlinear system:

$$
\begin{aligned}
A_R g_R\left(A_R^\top e^0, t_0\right) + A_V j_V^0 &= -A_C z_C^0 - A_L j_L^0 - A_I i_s(t_0) \\
A_V^\top e^0 &= v_s(t_0)
\end{aligned}
\tag{5.28}
$$

To obtain a solution (e^0, j_V^0) of (5.28) we apply Approach 4.25 without memristors. Then we compute the missing parts by:

$$
\begin{aligned}
z_L^0 &= A_L^\top e^0 \\
S_u M_\varepsilon^u G_u z_{\phi_u}^0 &= -\widetilde{S}_u M_\nu^u \widehat{a}_u^0 \\
M_\varepsilon^u z_{\widehat{a}_u}^0 &= -M_\varepsilon^u G_u z_{\phi_u}^0 + M_\sigma^u A_u A_E^\top e^0 - M_\sigma^u G_u \phi_u^0 - M_\sigma^u \widehat{\pi}_u^0 \\
z_{\widehat{\pi}_u}^0 &= \widehat{\pi}_u^0
\end{aligned}
$$

Remark 5.21. Due to the structure of the DAE (5.18) we obtain a locally unique solution through every consistent initial value and the perturbation index to be not greater than two, see Theorem 2.33, 2.46, 2.49 and 2.50.

5.4 Summary

In this chapter we have introduced circuits refined by spatially resolved electromagnetic devices and modeled by the modified nodal analysis and Maxwell's grid equations. The coupling is realized by the applied potential at the conductive contacts of the electromagnetic device and by the current through it. We discussed the structural properties of the coupled electromagnetic device/circuit system. The chosen coupling approach is different to [DHW04, Ben06, BBS11, Sch11], where the coupling is realized using serveral conductor models and applied as a source term.

We generalized the well-known topological index conditions of [Tis99, ET00] for the modified nodal analysis to circuits refined by spatially resolved electromagnetic devices modeled using Lorenz gauge (Theorem 5.15, 5.17 and 5.18) and proved that index-2 does not exceed. The index bound is also true for Coulomb gauge (Theorem 5.8 and 5.10). Furthermore we presented perturbation and solvability results for the coupled systems and the perturbation index does not exceed two (Remark 5.13 and 5.21). The electromagnetic devices were inserted into the circuit as controlled current sources, but the analysis showed that if using Lorenz gauge they, from the index point of view, did behave like inductances. We concluded that in case of an index-1 configuration it is always preferable to choose Lorenz gauge for the electromagnetic device.

Next, we presented an approach for the calculation of an operating point. Based on the linearity of the index-2 components (Lemma 5.12 and 5.20) the calculation of a consistent initialization is possible by correcting an operating point by solving a linear system. Due to the structure of the coupled system it is sufficient to start the numerical integration with the implicit Euler using an operating point to obtain a consistent initialization after the first time step.

6 Numerical Examples

In this chapter the different circuit models including memristors and electromagnetic devices are verified by some basic examples.

The simulation software is written in Python and is an extension in the framework of the MECS (Multiphysical Electric Circuit Simulator) developed by the group of Caren Tischendorf. The framework use for time integration a backward differentiation formulas implementation with an adaptive order and step size control for index-2 differential algebraic equations with a properly stated leading term which is based on [Tis96].

For the electromagnetic device simulation we integrate parts of the FIDES (Field Device Simulator) package of Sebastian Schöps, see [Sch11], implemented for the magnetoquasistatic device simulation. FIDES is written in OCTAVE and integrated within the framework of the demonstrator platform of the CoMSON project (Coupled Multiscale Simulation and Optimization in Nanoelectronics).

The 3D Visualizations are obtained by Paraview.

6.1 Index Behavior of Field Problems

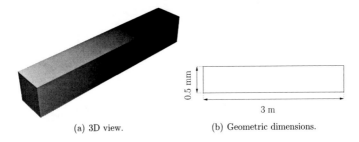

(a) 3D view.   (b) Geometric dimensions.

Figure 6.1: Geometry of the copper bar.

Let us consider a copper bar used in [BCS12] with a cross-sectional area of 0.25 mm^2 surrounded by air and discretized by the FIT, see Figure 6.1. The left contact is excited by a sinusoidal source of the form $v(t) = \sin(2\pi t)$, the other contact is grounded.

The simulations are carried out on the time interval $[0\,\text{s}, 0.5\,\text{s}]$ by the implicit Euler scheme with fixed step sizes $h = 8\text{e-5}$ s, 4e-5 s, 2e-5 s, 1e-5 s.

The numerical solution of the Lorenz (index-0) and Coulomb gauge (index-2) formulations of MGE are given in Figure 6.2(a) and 6.3(a). Both formulations provide solutions as anticipated.

To analyze the sensitivity of the formulations we perturb the sinusoidal source $v(t)$ by a small high-frequent noise

$$\delta(t) = 10^{-k}\sin(2 \cdot 10^{k+5}\pi t).$$

We get the perturbed source $v_p(t) = v(t) + \delta(t)$. For the simulation, in this thesis, we have chosen $k = 4$. As expected, the numerical solution of the perturbed Lorenz-based formulation is not affected, see Figure 6.2(b). On the other hand the solution of the index-2 formulation suffers strongly from the perturbation, see Figure 6.3(b). The effect

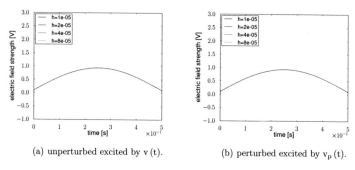

(a) unperturbed excited by $v(t)$.　　　(b) perturbed excited by $v_p(t)$.

Figure 6.2: Plot of a single component of the electric field using the Lorenz formulation (3.60).

occurs even for tiniest perturbations, that is, for very large $k \gg 1$. Moreover, the effect increases with a reduction in step size, that is, it cannot be compensated by a finer temporal mesh. This is a typical index-2 behavior: The error increases while the step

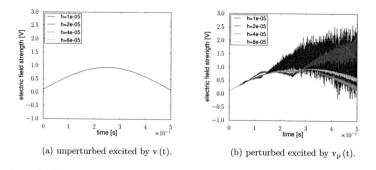

(a) unperturbed excited by $v(t)$.　　　(b) perturbed excited by $v_p(t)$.

Figure 6.3: Plot of a single component of the electric field using the Coulomb formulation (3.60).

size decreases. The index-2 problem is ill-conditioned but the perturbation error is not

136

propagated in time because the index-2 components enter only linearly. However, using a step size control we should exclude the index-2 variables for the step size prediction, because the numerical error might be detected by the step size control and leads to an unprofitable reduction of the step size. The best case would be an unreasonably small step size whereas in the worst case the integration could completely fail.

6.2 Memristive Circuits

(a) Basic memristor circuit.

(b) HP memristor circuit.

Figure 6.4: Memristor examples.

In this section we consider two models for the memristor to show that the MNA including memristor models (4.7) works properly, see Figure 6.4(a). The first example is the HP memristor stated in [SSSW08], see Figure 6.4(b). The HP device is composed of a

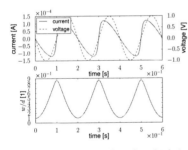

(a) Voltage and current through and relative boundary position of the device.

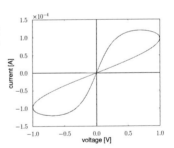

(b) Lissajous curve: pinched hysteresis.

Figure 6.5: HP memristor with R_{off} = 16e3 Ω and f = 5 Hz.

thin titanium dioxide film between two electrodes containing a doped (D) region and an undoped (U) region and thus it behaves as a semiconductor. The application of a voltage drop across the device moves the boundary between the two regions. With electric current passing in a given direction, the boundary between the two regions is moving in the same direction. The total device length is d and the length of the doped

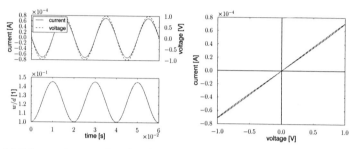

(a) Voltage and current through and relative boundary position of the device.

(b) Lissajous curve: pinched hysteresis.

Figure 6.6: HP memristor with $R_{off} = 16e3\ \Omega$ and $f = 50$ Hz.

region is denoted by $w \in [0, d]$. The limits of the memristor resistance is given by R_{off} and R_{on} for $w = 0$ and $w = d$. The dopant mobility is described by μ_V. The HP

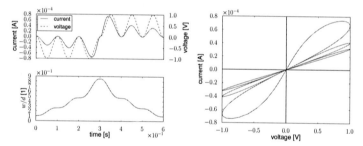

(a) Voltage and current through and relative boundary position of the device.

(b) Lissajous curve: pinched hysteresis.

Figure 6.7: HP memristor with $R_{off} = 36e3\ \Omega$ and $f = 5$ Hz.

memristor is modeled by the memristance

$$M(q, t) = R_{off} \left(1 - \frac{\mu_V R_{on}}{d^2} q \right) \tag{6.1}$$

with $R_{off} \gg R_{on}$ and the length of the doped region is given by

$$w = \frac{\mu_V R_{on}}{d} q.$$

The simulations were carried out on the time interval $[0\,\mathrm{s}, 0.6\,\mathrm{s}]$ by the implicit Euler scheme using the parameters $d = 1\text{e-}8$ m, $R_{\mathrm{on}} = 1\text{e}2\;\Omega$ and $\mu_V = 1\text{e-}13$ m/Vs. For Figure 6.5 and 6.6 we use as applied voltage source $v(t) = \sin(2\pi ft)$ and for Figure 6.7 the applied voltage source is given by

$$v(t) = \begin{cases} \sin(2\pi ft)^2, & \text{for } t \in [\,0, 0.3\,], \\ -\sin(2\pi ft)^2, & \text{for } t \in (0.3, 0.6]. \end{cases}$$

Unfortunately, the results shown in Figure 2 of [SSSW08] do not fit the stated parameters therein since the applied sinusoidal voltage source has in both cases a frequency of 5e-3 Hz instead of 1e2 Hz. Nonetheless the results show the same qualitative behavior as our results here.

In fact the HP memristance (6.1) is a polynomial. Another memristance described by a polynomial is given in [BBBK10] and reads

$$M(q, t) = r_1 + 3r_3 q^2$$

with $r_1 = 5$ V/A and $r_3 = 1\text{e}4$ V/A^{3}s^2. The results in [BBBK10] are obtained by the circuit given in Figure 6.4(a) in SPICE using a subcircuit to describe the behavior of the memristor.

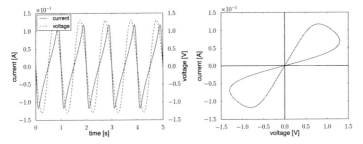

(a) Voltage and current through the device.   (b) Lissajous curve: pinched hysteresis.

Figure 6.8: Memristor with memristance described in [BBBK10].

Our simulations were carried out on the time interval $[0\,\mathrm{s}, 5\,\mathrm{s}]$ by the implicit Euler scheme and applied voltage source $v(t) = 13\text{e-}1\sin(2\pi t)$. The results are given in Figure 6.8 and fit perfectly to the simulation results given in [BBBK10].

6.3 Coupled Field/Circuit Problems

We regard two interlocking open copper loops with a cross-sectional area of 1 mm^2 surrounded by air and discretized by the FIT, see Figure 6.9(b). First we examine the basic circuit given in Figure 6.9(a).

The simulations were carried out for Coulomb gauge on the time interval $[0\,\text{s}, f^{-1}]$ by the implicit Euler scheme using as supply voltage sources $v(t) = 1\text{e-}3\sin(2\pi ft)$ with frequency f.

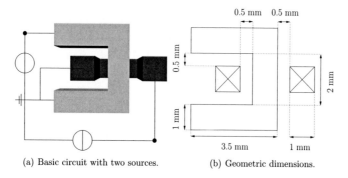

(a) Basic circuit with two sources. (b) Geometric dimensions.

Figure 6.9: Two interlocking open copper loops.

For $f = 1$ Hz we obtain the expected results since the static resistance of each open copper loop is between 124e-6 Ω and 158e-6 Ω. The results are given in Figure 6.10. For

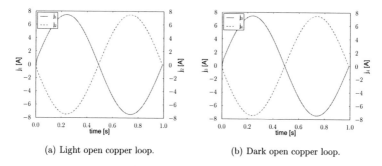

(a) Light open copper loop. (b) Dark open copper loop.

Figure 6.10: Current through the open copper loops with $f = 1$ Hz.

$f = 1\text{e}9$ Hz we obtain the results given in Figure 6.11. The current through the dark open copper loop is larger then the current through the light open copper loop. That behavior, of course, results from the increasing frequency and arises from the *proximity effect*, see [Ter43].

The effect can be described as follows: When an alternating electric current flows through an isolated conductor, it creates an associated alternating magnetic field around it, which influences the distribution by electromagnetic induction of an electric current flowing within an electrical conductor.

The alternating magnetic field induces eddy currents in adjacent conductors, altering the overall distribution of current flowing through them.

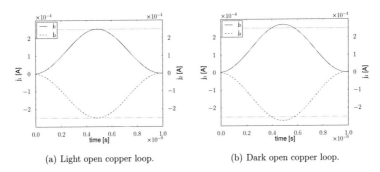

(a) Light open copper loop.　　　　(b) Dark open copper loop.

Figure 6.11: Two interlocking open copper loops with $f = $ 1e9 Hz.

Eddy currents are electric currents induced in conductors when a changing magnetic field acts on the conductor and causes a circulating flow of current within the conductor, see [Ter43]. These currents are responsible for the skin effect in conductors. The *skin effect* is the tendency of an alternating electric current to distribute itself within a conductor with the current density being largest near the surface of the conductor, decreasing at greater depths, that is, the electric current flows mainly at the skin of the conductor. This effect is due to opposing eddy currents induced by the changing magnetic field resulting from the alternating current. Figure 6.12 demonstrates well the increasing

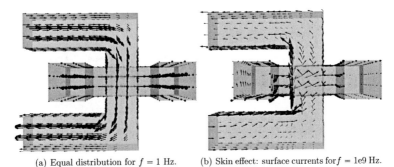

(a) Equal distribution for $f = $ 1 Hz.　　(b) Skin effect: surface currents for $f = $ 1e9 Hz.

Figure 6.12: Change in the distribution of the currents through the two interlocking open copper loops.

skin effect at increasing frequency for the basic circuit with two sources.

The second example circuit given in Figure 6.13 uses the two interlocking open copper loops of Figure 6.9(b), too. We choose $R_1 = 9865\text{e-}6\ \Omega$, $R_2 = 140\text{e-}6\ \Omega$, $L = 1\text{e-}2$ H

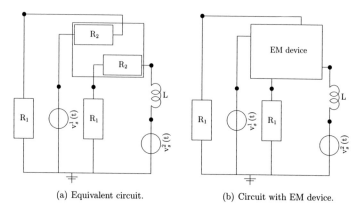

(a) Equivalent circuit.   (b) Circuit with EM device.

Figure 6.13: More complex circuit with EM device.

and the supply voltage sources to be $v(t) = 1\text{e-}1\sin(2\pi ft)$ with the frequency f. The simulations are carried out for Coulomb gauge on the time interval $[0\ \text{s}, 2f^{-1}]$ by the implicit Euler scheme. The results are given in Figure 6.14 and 6.15.

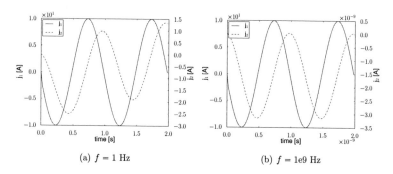

(a) $f = 1$ Hz   (b) $f = 1\text{e}9$ Hz

Figure 6.14: Equivalent circuit: Currents through the voltage sources.

In Figure 6.13(a) the EM device is replaced by two resistors with a resistance of R_2. For $f = 1$ Hz the results of both circuits are equal. This, of course, is not true for higher frequencies. There the inductive behavior of the EM device plays a crucial role. Amongst others, inductive coupling occurs and affects the circuit strongly, see above. Furthermore due to the skin effect the effective resistance of the device is increased.

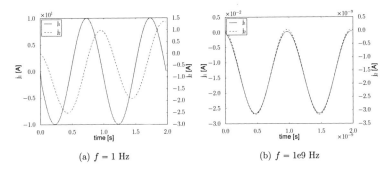

(a) $f = 1$ Hz
(b) $f = 1e9$ Hz

Figure 6.15: EM device circuit: Currents through the voltage sources.

Note that the equivalent circuit, Figure 6.13(a), is only for validating the low frequency results using the EM device.

6.4 Implementation Aspects

> The simplest model in applied mathematics is a system of linear equations. It is also by far the most important.
>
> GILBERT STRANG.

In every time integration step in transient simulation we have to solve linear systems. For effective solving of large linear systems iterative solvers and multigrid methods play an essential role since direct solvers are memory and time consuming and rapidly reach the limit of applicability. For an introduction to iterative solvers and multigrid methods we refer to [OST01, Saa03]. But most linear systems of the coupled simulation are not (directly) suitable for an iterative scheme since they are usually not symmetric, not positive definite and not diagonally dominant.

However, we often find the following structure of the systems:

$$Az = \begin{bmatrix} A_1 & A_2 \\ A_3 & A_4 \end{bmatrix} \begin{pmatrix} x \\ y \end{pmatrix} = \begin{pmatrix} u \\ v \end{pmatrix} \tag{6.2}$$

with $A_4 \in \mathbb{R}^{n_4 \times n_4}$, $A_1 \in \mathbb{R}^{n_1 \times n_1}$, $n_1 \ll n_4$ and A_4 being nonsingular and suitable for iterative solvers, while A_1 is not. Applying the Schur complement $S = \left(A_1 - A_2 A_4^{-1} A_3 \right)$ of the block A_4 we achieve the two linear systems

$$A_4 y = v - A_3 x \tag{6.3}$$

and

$$Sx = u - A_2 A_4^{-1} v. \tag{6.4}$$

The idea is to apply different linear solvers for solving the linear systems (6.3) and (6.4). To solve (6.4) we solve simultaneously $n_1 + 1$ linear systems of the dimension n_4 in advance, namely

$$A_4 w = v \tag{6.5}$$

and

$$A_4 W = A_3. \tag{6.6}$$

Then (6.4) becomes

$$(A_1 - A_2 W) x = u - A_2 w \tag{6.7}$$

For this, we solve (6.5) and (6.6) by an iterative solver. Then the linear system (6.7) can be solved by a direct solver to determine $x \in \mathbb{R}^{n_1}$.

Finally, we solve the linear system (6.3) by an iterative solver, too. We obtain the solution $y \in \mathbb{R}^{n_4}$ and thus we have solved the original linear system (6.2).

For the field/circuit system using Lorenz gauge (5.18) the Jacobian of the BDF methods for the time integration can be decomposed into the blocks are given by:

$$A_1 = \begin{bmatrix} \frac{\alpha_0}{h} A_C C \left(A_C^T e, t \right) A_C^T + A_R G \left(A_R^T e, t \right) A_R^T & A_L & A_V \\ -A_L^T & \frac{\alpha_0}{h} L \left(j_L, t \right) & 0 \\ A_V^T & 0 & 0 \end{bmatrix}$$

$$A_2 = \begin{bmatrix} A_E & 0 & 0 & 0 \\ & 0 & 0 & 0 \\ 0 & 0 & 0 & 0 \end{bmatrix}$$

$$A_3 = \begin{bmatrix} 0 & & 0 & 0 \\ 0 & & 0 & 0 \\ -\left(\frac{\alpha_0}{h} M_\varepsilon^u + M_\sigma^u \right) \Lambda_u A_E^T & 0 & 0 \\ 0 & & 0 & 0 \end{bmatrix}$$

$$A_4 = \begin{bmatrix} I & 0 & -\Lambda_u^T K_{\nu,d}^u (\widehat{a}_u) & 0 \\ 0 & \frac{\alpha_0}{h} \widetilde{S}_u M_\varepsilon^u G_u & \widetilde{S}_u M_\nu^u & 0 \\ 0 & \left(\frac{\alpha_0}{h} M_\varepsilon^u + M_\sigma^u \right) G_u & K_{\nu,d}^u (\widehat{a}_u) & \frac{\alpha_0}{h} M_\varepsilon^u + M_\sigma^u \\ 0 & 0 & \frac{\alpha_0}{h} I & -I \end{bmatrix}$$

Obviously, A_1 coming from the MNA (4.6) is not (directly) suitable for an iterative solver, whereas A_4 coming from the MGE (3.65) using Lorenz gauge could be suitable for iterative solvers, see Remark 3.41. This is motivated by the experience about the use of iterative solvers for MGE especially for MQS and electroquasistatic devices, see [Hip98, CSvW96, Cle98, DW04, Cle05, Sch11].

In [BCK^{+}11] we presented a systematic approach to reformulate the MNA (4.6) to be accessible for iterative solvers. Here the main goal was to eliminate the zero entries on the main diagonal of the Jacobian by manipulating the voltage sources. Nonetheless, this reformulation needs some effort and is usually not done within circuit simulation packages.

6.5 Summary

We have shown that the extended models work as expected due to our theoretical findings of the Chapters 3, 4 and 5. For the modified nodal analysis including memristors models we validated our model by using the HP memristor and an another memristance model from literature. We obtained the same qualitative results for both models

For the electromagnetic device we showed the influence of the chosen gauge with respect to perturbations and we observed the predicted behavior.

For the modified nodal analysis including electromagnetic device models we examined two simple circuits. In both cases we investigated two interlocking open copper loops with different frequencies for changing the applied potentials. For the low frequency case we observed the expected resistance behavior of the device. For the high frequency case we took note of the proximity and skin effect due to the inductive coupling.

The combination of methods for solving the resulting linear systems were briefly outlined.

Conclusion and Outlook

In this thesis, we presented the derivation of the modified nodal analysis including memristor models and coupled electromagnetic device/circuit models and we investigated the models in terms of the tractability index concept for differential-algebraic equations with a properly stated leading term.

We have derived new index results for circuits including memristors formulated as differential-algebraic equations with a properly stated leading term. We extended the well-known topological index conditions of [Tis99, ET00] for the modified nodal analysis to circuits including memristors. The critical index-2 circuit configurations are loops of only capacitors and voltage sources and cutsets of only inductors and current sources. We concluded that, from index point of view, the memristors behave like resistors.

The electromagnetic devices were modeled by Maxwell's equation in a potential formulation using the finite integration technique for the resulting spatial discretization. General properties of the discrete operators have been discussed. The spatial discretization leads to Maxwell's grid equations, which were formulated in terms of potentials with incorporated boundary conditions using a new class of discrete gauge conditions in terms of the finite integration technique based on Lorenz gauge. The structural properties of Maxwell's grid equations were discussed and analyzed by the index concept to obtain knowledge about the stability of the solutions with respect to perturbations. The main result here is that the index depends on the chosen gauge condition. For Coulomb gauge we obtain an index-2 differential-algebraic equation with a properly stated leading term whereas for Lorenz gauge we achieve an ordinary differential equation.

The coupled electromagnetic device/circuit system was formulated as a differential-algebraic equation with a properly stated leading term. The coupling was realized by the applied potential at the conductive contacts of the electromagnetic device and by the current through it. In our case we have taken a different coupling approach than [DHW04, Ben06, BBS11, Sch11], where the coupling is realized using several conductor models and applied as a source term. We generalized the well-known topological index conditions for the modified nodal analysis to circuits refined by spatially resolved electromagnetic devices modeled in case of using Lorenz gauge. In case of Lorenz gauge the critical index-2 circuit configurations are loops of only capacitors and voltage sources and cutsets of only inductors, electromagnetic devices and current sources. For Coulomb gauge we have shown that the index does not exceed two. The electromagnetic devices were inserted into the circuit as controlled current sources, but the analysis showed that when applying Lorenz gauge they, from the index point of view, behave like inductances.

We concluded that in case of an index-1 circuit configuration it is always preferable to choose Lorenz gauge for the electromagnetic device.

All considered differential-algebraic equations resulting from our fields of applications have a common structure such as a properly stated leading term, constant projectors onto/along certain subspaces and linear index-2 variables. For that reason we investigated the differential-algebraic equations in a common differential-algebraic equation framework. For index-2 differential-algebraic equations we derived a local uniquely solvability result and a perturbation estimation. To achieve these results we extended the well-known index reduction techniques for differential-algebraic equations without a properly stated leading term to differential-algebraic equations with a properly stated leading term and exploited local uniquely solvability and perturbation results for index-1 differential-algebraic equations with a properly stated leading term from literature. In particular we have proved that if the differential-algebraic equations with a properly stated leading term have index-2 then the differential-algebraic equation has perturbation index-2 as well. We extended the step-by-step approach by [Est00] for calculating consistent initial values for differential-algebraic equations with a properly stated leading term using the linearity of the index-2 components. In addition we presented an approach to calculate an operating point.

Some numerical examples were given to show that the models works as expected.

There are still a lot of unsolved problems and tasks to be tackled. For electromagnetic devices further sets of consistent boundary conditions for the coupled electromagnetic device/circuit model and other models for nonlinear materials have to be considered. Beyond that it would be of great interest to combine existing models for electromagnetic devices, semiconductor devices and heating of devices. For this the next step could be to combine electromagnetic and semiconductor device models to study the influence of electromagnetic fields of the surrounding circuitry to semiconductor switching elements. Then an index analysis of the coupled electromagnetic-semiconductor device/circuit model would be an inevitable future step for the numerical simulation. Finally, heat effects could be studied by extending the models by heat conducting model equations.

A Linear Algebra

We make use of several simple definitions and deductions from linear algebra in this thesis and collect the results in this Chapter.

Lemma A.1. Let be $A \in \mathbb{R}^{n \times m}$. Then

$$\ker A = \left(\operatorname{im} A^\top \right)^\perp \text{ and } \operatorname{im} A = \left(\ker A^\top \right)^\perp$$

holds true.

Proof. We show both inclusions.

($\subseteq$) Let be $x \in \ker A$. Then

$$A^\top z \in \operatorname{im} A^\top \Rightarrow 0 = (Ax)^\top z = x^\top A^\top z \Rightarrow x \in \left(\operatorname{im} A^\top \right)^\perp$$

is true. For $z \in \ker A^\top$ we obtain

$$Ax \in \operatorname{im} A \Rightarrow 0 = x^\top \left(A^\top z \right) = (Ax)^\top z \Rightarrow z \in \left(\ker A^\top \right)^\perp.$$

($\supseteq$) Starting with

$$\dim \left(\operatorname{im} A^\top \right)^\perp \geqslant \dim \ker A = n - \dim \operatorname{im} A \geqslant n - \dim \left(\ker A^\top \right)^\perp$$
$$= n - \left(m - \dim \ker A^\top \right) = n - \dim \operatorname{im} A^\top = \dim \left(\operatorname{im} A^\top \right)^\perp$$

we get

$$\dim \ker A = \dim \left(\operatorname{im} A^\top \right)^\perp \text{ and } \dim \operatorname{im} A = \dim \left(\ker A^\top \right)^\perp.$$

$\square$

Definition A.2. A (non-symmetric) matrix $A \in \mathbb{R}^{n \times n}$ is called *positive semidefinite* if and only if $x^\top Ax \geqslant 0$ and *positive definite* if and only if $x^\top Ax > 0$ for all $x \neq 0$, $x \in \mathbb{R}^n$.

From the definition above several simple results can be elementarily derived.

Lemma A.3. Let $A \in \mathbb{R}^{m \times m}$ be positive definite and $B \in \mathbb{R}^{k \times m}$. Then

$$\ker BAB^\top = \ker B^\top \text{ and } \operatorname{im} BAB^\top = \operatorname{im} B$$

holds true.

Proof. At first we show $\ker BAB^\top = \ker B^\top$.

($\subseteq$) Let be $x \in \ker BAB^\top$. Then

$$x^\top BAB^\top x = 0 \Rightarrow y^\top Ay = 0, \ y = B^\top x \Rightarrow B^\top x = 0 \Rightarrow x \in \ker B^\top,$$

since A is positive definite.

($\supseteq$) If $x \in \ker B^\top$ then $BAB^\top x = 0$ and hence $x \in \ker BAB^\top$.

Due to Lemma A.1 and we have

$$\operatorname{im} BAB^\top = \left(\ker BA^\top B^\top\right)^\perp = \left(\ker B^\top\right)^\perp = \operatorname{im} B,$$

since $A^\top$ is positive definite. $\qquad\qquad\qquad\qquad\qquad\qquad\qquad\qquad$ $\square$

A.1 Properties of Projectors

> Für euch ist es einfach. Ihr seid mit Projektoren aufgewachsen.
>
> ROSWITHA MÄRZ DURING THE "DEUTSCHE MATHEMATIKER-VEREINIGUNG" CONFERENCE, SEPT. 20, 2011, COLOGNE.

This section is devoted to basic definitions and results in projector calculus.

Definition A.4. The basics definitions for a projector are:

(i) A matrix $Q \in \mathbb{R}^{m \times m}$ is called *projector*, if $Q^2 = Q$.

(ii) A projector $Q \in \mathbb{R}^{m \times m}$ is called *projector onto* $\mathcal{S} \subseteq \mathbb{R}^m$, if $\operatorname{im} Q = \mathcal{S}$.

(iii) A projector $Q \in \mathbb{R}^{m \times m}$ is called *projector along* $\mathcal{S} \subseteq \mathbb{R}^m$, if $\ker Q = \mathcal{S}$.

(iv) A projector $Q \in \mathbb{R}^{m \times m}$ is called *orthogonal projector* if $Q = Q^\top$.

Lemma A.5. Let $Q \in \mathbb{R}^{m \times m}$ be a projector and $P = I - Q$. Then:

(i) P is a projector.

(ii) $x \in \operatorname{im} Q \Leftrightarrow x = Qx$.

(iii) $\ker P = \operatorname{im} Q$ and $\ker Q = \operatorname{im} P$.

(iv) $\ker Q \oplus \operatorname{im} Q = \mathbb{R}^m$.

(v) If Q is nonsingular, then $Q = I$.

(vi) Let $\overline{Q}$ be a projector with $\overline{P} = I - \overline{Q}$ and $\operatorname{im} Q = \operatorname{im} \overline{Q}$. Then $Q\overline{Q} = \overline{Q}$ and $P\overline{P} = P$ hold true.

(vii) $I + PEQ$ is nonsingular for all $E \in \mathbb{R}^{m \times m}$.

Proof. The proofs are straightforward.

(i) $Q^2 = Q \Leftrightarrow (I - P)^2 = I - P \Leftrightarrow P^2 = P$.

(ii) $x \in \operatorname{im} Q \Leftrightarrow \exists z \in \mathbb{R}^m : x = Qz$ and $Qx = Qz \Leftrightarrow x = Qx$.

(iii) $x \in \ker P \Rightarrow Px = 0 \Rightarrow x = Qx \Rightarrow x \in \operatorname{im} Q$. The other one is analog.

(iv) $x = Qx + Px,\ x \in \ker Q \cap \operatorname{im} Q \Rightarrow Qx = 0, x = Qx \Rightarrow x = 0$.

(v) $Q^2 = Q \Leftrightarrow Q^{-1}Q^2 = I \Leftrightarrow Q = I$.

(vi) Let be $x \in \mathbb{R}^m$. Then $\overline{Q}x \in \operatorname{im} \overline{Q} = \operatorname{im} Q$. Hence $\overline{Q}x = Q\overline{Q}x$ and $\overline{Q} = Q\overline{Q}$. We get

$$P\overline{P} = (I - Q)\left(I - \overline{Q}\right) = I - Q - \overline{Q} - Q\overline{Q} = I - Q = P.$$

(vii) The inverse is given by $I - PEQ$.

$\square$

If Q is a projector then we call $P = I - Q$ the *complementary projector*.

Remark A.6. The product of two projectors is not necessarily a projector, too. For that we look at the projectors P_1 and P_2 given by

$$P_1 = \begin{bmatrix} 1 & 1 \\ 0 & 0 \end{bmatrix} \text{ and } P_2 = \begin{bmatrix} 1 & 0 \\ 1 & 0 \end{bmatrix}.$$

Obviously $P_3 = P_1 P_2$ is not a projector.

Lemma A.7. Let $A \in \mathbb{R}^{m \times n}$ and $Q \in \mathbb{R}^{m \times m}$ be a projector onto $\ker A^\top$. Then

$$\ker Q^\top = \operatorname{im} A$$

holds true.

Proof. Using Lemma A.1 we obtain

$$\ker Q^\top = (\operatorname{im} Q)^\perp = \left(\ker A^\top\right)^\perp = \operatorname{im} A.$$

$\square$

Remark A.8. Instead of determining a projector Q along $\operatorname{im} A$ we can calculate $Q^\top$ onto $\ker A^\top$ and betimes the computation of a projector onto a subspace is easier.

Lemma A.9. Let be $A \in \mathbb{R}^{n \times n}$. Then:

(i) $Q \in \mathbb{R}^{n \times n}$ a projector onto $\operatorname{im} A$ implies $QA = A$.

(ii) $Q \in \mathbb{R}^{n \times n}$ a projector along $\ker A$ implies $AQ = A$.

Proof. We get:

(i) $\operatorname{im} Q = \ker (I - Q) = \operatorname{im} A \Rightarrow (I - Q) A = 0 \Rightarrow QA = A$

(ii) $\ker Q = \operatorname{im} (I - Q) = \ker A \Rightarrow A (I - Q) = 0 \Rightarrow AQ = A$

$\square$

The next Lemma is motivated by [ET00].

Lemma A.10. Let $A \in \mathbb{R}^{m \times m}$ be positive definite, $B \in \mathbb{R}^{k \times m}$, Q be a projector onto $\ker B^\mathsf{T}$ and $P = I - Q$. Then the matrix

$$H = BAB^\mathsf{T} + Q^\mathsf{T}Q$$

is positive definite and

$$HP = BAB^\mathsf{T} = P^\mathsf{T}H$$

holds true.

Proof. It is clear that H is positive semidefinite since it is the sum of two positive semidefinite matrices. With that we get

$$z^\mathsf{T}Hz = 0 \Leftrightarrow \begin{cases} z^\mathsf{T}BAB^\mathsf{T}z = 0 \\ z^\mathsf{T}Q^\mathsf{T}Qz = 0 \end{cases}$$

and we obtain $z = 0$ by reason of $z \in \ker B^\mathsf{T} = \operatorname{im} Q$ and $z \in \ker Q$. Hence H is positive definite. The second statement follows immediately:

$$HP = \left(BAB^\mathsf{T} + Q^\mathsf{T}Q\right) P = BAB^\mathsf{T} = P^\mathsf{T} \left(BAB^\mathsf{T} + Q^\mathsf{T}Q\right) = P^\mathsf{T}H$$

$\square$

For computational aspects of projectors calculus we refer to [LMT13], where the projectors are determined by matrix decompositions.

A.2 Generalized Inverse

We report the (basic) definitions and relations of generalized inverses needed for our considerations. A more detailed look on this topic is provided by, for example, [BIG03, BO71].

Definition A.11. With $A^- \in \mathbb{R}^{n \times m}$ we denote a *pseudoinverse* and $\{1,2\}$-inverse of $A \in \mathbb{R}^{m \times n}$ if

$$AA^-A = A \text{ and } A^-AA^- = A^-$$

are fulfilled.

Lemma A.12. A pseudoinverse $A^- \in \mathbb{R}^{n \times m}$ exists for every $A \in \mathbb{R}^{m \times n}$.

Proof. For every $A \in \mathbb{R}^{m \times n}$ it exist nonsingular matrices $S \in \mathbb{R}^{m \times m}$ and $T \in \mathbb{R}^{n \times n}$ with

$$SAT = \begin{bmatrix} I & 0 \\ 0 & 0 \end{bmatrix} \Leftrightarrow A = S^{-1} \begin{bmatrix} I & 0 \\ 0 & 0 \end{bmatrix} T^{-1}$$

with $I \in \mathbb{R}^{r \times r}$ and $r = \operatorname{rank} A$. The matrix

$$A^- = T \begin{bmatrix} I & X \\ Y & YX \end{bmatrix} S$$

fulfills all necessary properties, where $Y \in \mathbb{R}^{(m-r) \times r}$, $X \in \mathbb{R}^{r \times (n-r)}$ are arbitrarily, see [BO71]. $\qquad \square$

Lemma A.13. The matrices $AA^- \in \mathbb{R}^{m \times m}$ and $A^-A \in \mathbb{R}^{n \times n}$ are projectors onto $\operatorname{im} A$ and along $\ker A$.

Proof. The projector properties are clear due to the definition of the pseudoinverse. It remains to show $\operatorname{im} AA^- = \operatorname{im} A$ and $\ker A^-A = \ker A$. We show the first identity. We have:

($\subseteq$) $x \in \operatorname{im} AA^- \Rightarrow x = AA^-z \Rightarrow x = Ay \Rightarrow x \in \operatorname{im} A$

($\supseteq$) $x \in \operatorname{im} A \Rightarrow x = Az \Rightarrow x = AA^-Az \Rightarrow x = AA^-y \Rightarrow x \in \operatorname{im} AA^-$

The second identity is completely analog. $\qquad \square$

Let $R \in \mathbb{R}^{n \times n}$ be a projector onto $\operatorname{im} A$ and $P \in \mathbb{R}^{m \times m}$ a projector along $\ker A$.

Theorem A.14. Let $R \in \mathbb{R}^{n \times n}$ be a projector onto $\operatorname{im} A$ and $P \in \mathbb{R}^{m \times m}$ a projector along $\ker A$. The choice

(i) $A^-A = P$

(ii) $AA^- = R$

is always possible.

153

Proof. Let be $I \in \mathbb{R}^{r \times r}$ and $r = \operatorname{rank} A$. It exists nonsingular matrices $S \in \mathbb{R}^{m \times m}$ and $T \in \mathbb{R}^{n \times n}$ with

$$A^- = T \begin{bmatrix} I & X \\ Y & YX \end{bmatrix} S, \ A^- A = T \begin{bmatrix} I & 0 \\ Y & 0 \end{bmatrix} T^{-1}, \text{ and } AA^- = S^{-1} \begin{bmatrix} I & X \\ 0 & 0 \end{bmatrix} S,$$

see proof of Lemma A.12.

(i) Let be $z \in \ker P = \ker A^- A$. Then $y \in \ker P_1 = \ker P_2$ with

$$P_1 = T^{-1} P T, \ P_2 = \begin{bmatrix} I & 0 \\ Y & 0 \end{bmatrix} \text{ and } y = T^{-1} z.$$

The matrices P_1 and P_2 are projectors with the same nullspace. We have

$$P_2 y = 0 \Rightarrow (0, y_2) \in \ker P_2 \text{ with } y = (y_1, y_2)$$

and hence we obtain

$$P_1 = \begin{bmatrix} P_{1,1} & 0 \\ P_{1,2} & 0 \end{bmatrix} \text{ and } P_1^2 = \begin{bmatrix} P_{1,1}^2 & 0 \\ P_{1,2}P_{1,1} & 0 \end{bmatrix}.$$

Since P_1 is a projector we conclude that $P_{1,1}$ is a projector, too, and $P_{1,2} = P_{1,2}P_{1,1}$. Assume $P_{1,1} \neq I$, that is, $P_{1,1}$ is singular. Then there exist $y_1 \neq 0$ with $P_{1,1}y_1 = 0$ and $P_1 y = 0$ with $y = (y_1, y_2)$. With $y_1 \neq 0$ we obtain

$$P_2 y = \begin{pmatrix} y_1 \\ Y y_1 \end{pmatrix} \neq \begin{pmatrix} 0 \\ 0 \end{pmatrix}$$

which is a contradiction to the property that both nullspaces equals. Hence $P_{1,1} = I$ and

$$T^{-1} P T = \begin{bmatrix} I & 0 \\ P_{1,2} & 0 \end{bmatrix}$$

with $Y = P_{1,2}$.

(ii) Let be $z \in \operatorname{im} R = \operatorname{im} AA^-$. Then $y \in \operatorname{im} R_1 = \operatorname{im} R_2$ with

$$R_1 = S R S^{-1}, \ R_2 = \begin{bmatrix} I & X \\ 0 & 0 \end{bmatrix} \text{ and } y = S z.$$

The matrices R_1 and R_2 are projectors with the same image. We have

$$y \in \operatorname{im} R_2 \Rightarrow y = (y_1, 0) \text{ with } y = (y_1, y_2)$$

and hence we obtain

$$R_1 = \begin{bmatrix} R_{1,1} & R_{1,2} \\ 0 & 0 \end{bmatrix} \text{ and } R_1^2 = \begin{bmatrix} R_{1,1}^2 & R_{1,1}R_{1,2} \\ 0 & 0 \end{bmatrix}.$$

Since R_1 is a projector we conclude that $R_{1,1}$ is a projector, too, and $R_{1,2} = R_{1,1}R_{1,2}$. Assume $R_{1,1} \neq I$, that is, $R_{1,1}$ is singular. Then there exist $y_1 \neq 0$, $y_1 \in \mathbb{R}^r$, with $R_{1,1}y_1 = 0$. Thus $y_1 \notin \text{im } R_{1,1}$ and hence $(y_1, 0) \notin \text{im } R_1$. But $(y_1, 0) \in \text{im } R_2$ since

$$\begin{pmatrix} y_1 \\ 0 \end{pmatrix} = \begin{bmatrix} I & X \\ 0 & 0 \end{bmatrix} \begin{pmatrix} y_1 \\ 0 \end{pmatrix}.$$

That is a contradiction to the property that both images equals. Hence $R_{1,1} = I$ and

$$SRS^{-1} = \begin{bmatrix} I & R_{1,2} \\ 0 & 0 \end{bmatrix}$$

with $X = R_{1,2}$. $\qquad\square$

A proof of Theorem A.14 is already given in [BIG03], Chapter 2, Theorem 12. But the proof given above provides a way to construct the pseudoinverse A^- explicitly with the special choice above.

Lemma A.15. Let $R \in \mathbb{R}^{n \times n}$ be a projector onto $\text{im } A$ and $P \in \mathbb{R}^{m \times m}$ a projector along $\ker A$. The pseudoinverse A^- together with

$$A^-A = P \text{ and } AA^- = R$$

is uniquely determine.

Proof. Let B be a pseudoinverse of A too. Then:

$$B = BAB = BR = BAA^- = PA^- = A^-AA^- = A^-$$

$\qquad\square$

Definition A.16. With $A^+ \in \mathbb{R}^{n \times m}$ we denote the *Moore-Penrose pseudoinverse* and $\{1, 2, 3, 4\}$-inverse of $A \in \mathbb{R}^{m \times n}$ if

$$AA^+A = A \qquad\qquad A^+AA^+ = A^+$$
$$A^+A = \left(A^+A\right)^\top \qquad\qquad AA^+ = \left(AA^+\right)^\top$$

are fulfilled.

The Moore-Penrose pseudoinverse is a special pseudoinverse. In contrast to a pseudoinverse we require also that A^-A and AA^- are orthogonal projectors

Lemma A.17. A Moore-Penrose pseudoinverse A^+ exist for every $A \in \mathbb{R}^{m \times n}$.

Proof. Let $A = U\Sigma V^\top$ be a singular value decomposition. The Moore-Penrose pseudoinverse is given by

$$A^+ = V \begin{bmatrix} D^{-1} & 0 \\ 0 & 0 \end{bmatrix} U^\top \text{ and } \Sigma = \begin{bmatrix} D & 0 \\ 0 & 0 \end{bmatrix}.$$

The necessary properties follow immediately. $\qquad\square$

An essential difference between the pseudoinverse A^- and the Moore-Penrose pseudoinverse A^+ is the uniqueness of the latter.

Lemma A.18. The Moore-Penrose pseudoinverse A^+ of $A \in \mathbb{R}^{m \times n}$ is uniquely determine.

Proof. Let B be also Moore-Penrose pseudoinverse of A. Then

$$\begin{aligned} B &= BAB = B\,(AB)^\top = BB^\top A^\top = BB^\top \left(AA^+A\right)^\top = BB^\top A^\top \left(A^+A\right)^\top \\ &= B\,(AB)^\top \left(A^+A\right)^\top = BABAA^+ = BAA^+ = (BA)^\top A^+AA^+ \\ &= (BA)^\top \left(A^+A\right)^\top A^+ = A^\top B^\top A^\top \left(A^+\right)^\top A^+ = (ABA)^\top \left(A^+\right)^\top A^+ \\ &= A^\top \left(A^+\right)^\top A^+ = \left(A^+A\right)^\top A^+ = A^+AA^+ = A^+ \end{aligned}$$

is valid. □

Lemma A.19. The matrices $AA^+ \in \mathbb{R}^{m \times m}$ and $A^+A \in \mathbb{R}^{n \times n}$ are projectors along $\ker A^\top$ and onto $\operatorname{im} A^\top$.

Proof. The projector properties is clear due to the definition of the pseudoinverse. It remains to show $\ker AA^+ = \ker A^\top$ and $\operatorname{im} A^+A = \operatorname{im} A^\top$. We show the first identity. We have

$$\ker AA^+ = \ker \left(AA^+\right)^\top = \left(\operatorname{im} AA^+\right)^\perp = (\operatorname{im} A)^\perp = \ker A^\top.$$

The second identity is completely analog. □

Remark A.20. If $A \in \mathbb{R}^{n \times n}$ is nonsingular, the pseudoinverse and the inverse coincide.

Lemma A.21. Let $A \in \mathbb{R}^{m \times m}$ be positive definite and $B \in \mathbb{R}^{k \times m}$. Then

$$\operatorname{im} B^+B = \operatorname{im} B^+BAB^\top$$

holds true.

Proof. We show both inclusions.

($\subset$) Let be $x \in \operatorname{im} B^+B$. Then there exists $y \in \mathbb{R}^m$ such that $x = B^+By$ and $z \in \operatorname{im} B$ with $z = By$. Lemma A.3 leads to $z \in \operatorname{im} BAB^\top$ and there exists $u \in \mathbb{R}^k$ such that $z = BAB^\top u$. Hence we obtain $x = B^+BAB^\top u$ and $x \in \operatorname{im} B^+BAB^\top$.

($\supset$) Let be $x \in \operatorname{im} B^+BAB^\top$. Then there exists $y \in \mathbb{R}^k$ such that $x = B^+BAB^\top y$. Choosing $z = AB^\top y$ we obtain $x = B^+Bz$ and $x \in \operatorname{im} B^+B$.

□

For computational aspects of generalized inverses calculus we refer to [LMT13], where the generalized inverses a determine by matrix decompositions.

B Graph Theory

In this section we want to introduce elementary basics derived from the theory of graphs and digraphs. For more details we refer the reader to [Die05]. First, we start with some basic notation and definitions. Roughly speaking, a graph is a set of edges and the ends of the edges are called nodes. If all edges own an orientation then the graph is called a digraph. Let $\mathcal{N}$ be a set. Then $|\mathcal{N}| \in \mathbb{N}$ is the number of elements in $\mathcal{N}$.

Definition B.1 (graph, node, edges). A *graph* $\mathcal{G}$ is a tuple of finite sets $\mathcal{G} := (\mathcal{N}, \mathcal{E})$ such that $\mathcal{E} \subseteq \mathcal{N} \times \mathcal{N}$ with $|\mathcal{N}|, |\mathcal{E}| < \infty$. We call an element of the set $\mathcal{N}$ *node* and of the set $\mathcal{E}$ *edge*. In general each edge corresponds to an unsorted tuple of nodes denoted by $e = (n_1, n_2)$ and $e = (n_2, n_1)$, respectively, with $e \in \mathcal{E}$ and $n_1, n_2 \in \mathcal{N}$. We call n a node of $\mathcal{G}$ if $n \in \mathcal{N}$ and e an edge of $\mathcal{G}$ if $e \in \mathcal{E}$.

A graph, where edges correspond to an unsorted tuple of nodes, is called *undirected graph*. We say that two nodes n_1 and n_2 of $\mathcal{G}$ are adjacent if either $e = (n_1, n_2)$ or $e = (n_2, n_1)$ are edges of $\mathcal{G}$. In case of $e = (n_1, n_2)$ we say that n_1 is the front node and n_2 is the back node of edge e. A node n of $\mathcal{G}$ is called incident to an edge e of $\mathcal{G}$ if n is the front or back node of e. Two edges e_1 and e_2 of $\mathcal{G}$ are called incident if these edges have one common node n of $\mathcal{G}$ and an edge e of $\mathcal{G}$ is called incident with a node n of $\mathcal{G}$ if the node n is the front or back node of the edge e .

A common approach to illustrate a graph is drawing a dot for each node and joining two of dots by a line if there exists an edge between these two dots. How to draw the dots and edges is considered irrelevant because all relevant information is the node-to-edge-relation. Hence the representation of a graph is not unique.

For further investigations we exclude the possibility to have an edge with the same front and back node.

Definition B.2 (digraph). A *digraph* $\mathcal{G}$ is a graph, where each edge corresponds to a sorted tuple of nodes. We say that the edges are orientated.

In case of digraphs the representation of edges are considered as arrows instead of lines. Each digraph $\mathcal{G}$ we can assign an undirected graph by dropping the edge orientation. If we speak about graphs we include digraphs by the assigned undirected graph.

Definition B.3 (path). A set of n edges $\{e_1, \ldots, e_n\} \subseteq \mathcal{E}$ of a graph $\mathcal{G}$ is called a *path* between n_1 and n_2 if:

(i) the edges e_i and e_{i+1} are incident, $i \in \{1, \ldots, n-1\}$

(ii) each node is incident to at most two edges

(iii) the nodes n_1, n_2 belong to exactly one edge of the set

Definition B.4 (Connected graph). A graph is called a *connected graph* if there exists at least one path between any two nodes of the graph. Otherwise we call the graph disconnected.

For defining loops, trees and cutsets of a graph we need the following definition of a subgraph.

Definition B.5 (subgraph). A graph $\mathcal{G}' := (\mathcal{N}', \mathcal{E}')$ is called a *subgraph* of $\mathcal{G}$ if $\mathcal{N}' \subseteq \mathcal{N}$, $\mathcal{E}' \subseteq \mathcal{E}$ and $\mathcal{E}' \subseteq \mathcal{N}' \times \mathcal{N}'$. The difference graph $\mathcal{G} \backslash \mathcal{G}'$ is given by $\mathcal{G} \backslash \mathcal{G}' = (\mathcal{N}, \mathcal{E} \backslash \mathcal{E}')$ and for $e \in \mathcal{E}$ we define $(\mathcal{G} \backslash \mathcal{G}') \cup \{e\} = (\mathcal{N}, (\mathcal{E} \backslash \mathcal{E}') \cup \{e\})$.

Next we can define loops, trees and cutsets of a graph.

Definition B.6 (loop). A subgraph $\mathcal{G}'$ of a connected graph $\mathcal{G}$ is called a *loop* if it is connected and precisely two edges of it are incident with each node.

Definition B.7 (tree). A subgraph $\mathcal{G}'$ of a connected graph $\mathcal{G}$ is called a *tree* if:

(i) $\mathcal{G}'$ is connected

(ii) $\mathcal{G}'$ contains all nodes of $\mathcal{G}$

(iii) $\mathcal{G}'$ has no loops

It should be mentioned that we can construct a tree for each connected graph. Furthermore each tree of a connected graph with $|\mathcal{N}|$ nodes consists of exactly $|\mathcal{N}| - 1$ edges, see [Die05] Proposition 1.5.3 and 1.5.6.

Definition B.8 (cutset). A subgraph $\mathcal{G}'$ of a connected graph $\mathcal{G}$ is called a *cutset* if:

(i) $\mathcal{G} \backslash \mathcal{G}'$ is disconnected

(ii) For every $e' \in \mathcal{E}'$ the graph $(\mathcal{G} \backslash \mathcal{G}') \cup \{e'\}$ is connected

Now we will focus on digraphs and combine some linear algebra with graph theory by introducing the so-called incidence matrix for digraphs. We obtain a matrix representation for a graph, which shows the relationship between nodes and edges.

Definition B.9 (incidence matrix). Let a digraph $\mathcal{G}$ with $|\mathcal{N}|$ nodes and $|\mathcal{E}|$ edges be given. The *incidence matrix* denoted by $A_a \in \{-1, 0, 1\}^{|\mathcal{N}| \times |\mathcal{E}|}$ is defined as $A_a = (a_{ij})$ with

$$a_{ij} = \begin{cases} 1 & \text{if the edge } j \text{ leaves the node } i, \\ -1 & \text{if the edge } j \text{ enters the node } i, \\ 0 & \text{else.} \end{cases}$$

By definition of the incidence matrix A_a of a connected digraph it is easy to see that the rows are linear dependent. To be more specific the sum of all rows of A_a equals zero. This is caused by the fact that each column contains exactly one 1 and one -1 and all other entries are zero. This becomes obvious if one observes that each column corresponds to exactly one edge and each edge has two incident nodes. Hence one row of the incidence matrix can be neglected in order to describe the graph. That node corresponding to the neglected row is called *reference node* and can be chosen freely. Erasing one row of A_a we obtain the so-called *reduced incidence matrix* A. In literature the reduced incidence matrix A is often called only incidence matrix A. In our case we will name the matrix A incidence matrix too.

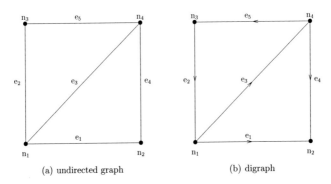

Figure B.1: An undirected graph and digraph with four nodes and five edges.

Example B.10. Regarding the graph $\mathcal{G} = (\mathcal{N}, \mathcal{E})$ in Figure B.1(a) the set of nodes is given by $\mathcal{N} = \{n_1, n_2, n_3, n_4\}$ and the set of edges by $\mathcal{E} = \{e_1, e_2, e_3, e_4, e_5\}$. The graph $\mathcal{G}$ is the undirected version of the digraph in Figure B.1(b). Obviously the graph $\mathcal{G}$ is connected. A loop is given by the edges $\{e_1, e_4, e_3\}$, but not by the edges $\{e_2, e_5\}$. The last set of edges builds a path starting in node n_1 and ending in node n_4. In this case the set describes a cutset, too. A tree is given by the edges $\{e_1, e_3, e_5\}$. Choosing the node n_4 as reference node the incidence matrices are

$$
A_a = \begin{bmatrix} 1 & -1 & 1 & 0 & 0 \\ -1 & 0 & 0 & -1 & 0 \\ 0 & 1 & 0 & 0 & -1 \\ 0 & 0 & -1 & 1 & 1 \end{bmatrix} \text{ and } A = \begin{bmatrix} 1 & -1 & 1 & 0 & 0 \\ -1 & 0 & 0 & -1 & 0 \\ 0 & 1 & 0 & 0 & -1 \end{bmatrix}
$$

respectively.

In the following we collect some facts about incidence matrices.

Theorem B.11. The incidence matrix A of a connected graph $\mathcal{G}$ with $|\mathcal{N}|$ nodes has $|\mathcal{N}| - 1$ linear independent rows.

Theorem B.12. A subgraph $\mathcal{G}'$ of a connected graph $\mathcal{G}$ with $|\mathcal{E}'|$ edges has loops if and only if the columns of the incidence matrix A corresponding to these $|\mathcal{E}'|$ edges are linear dependent.

Theorem B.13. Let A be the incidence matrix of a connected graph $\mathcal{G}$ with $|\mathcal{N}|$ nodes. Then $|\mathcal{N}| - 1$ columns of A are linear independent if and only if the edges of these columns form a tree.

We refer the reader to Appendix A.1 in [Tis04] for the proofs.

Remark B.14 (Incidence matrices). We note the following:

(i) An incidence matrix A_X of a subgraph has full column rank if and only if there is no loop (in the subgraph), that is, no X-loop, see Theorem B.12.

(ii) Let $\begin{bmatrix} A_X & A_Y \end{bmatrix}$ denote the incidence matrix of a connected graph. Then $A_X^\top$ has full column rank if and only if there is a spanning tree of elements from A_X, that is, no Y-cutset, see Theorem B.13.

C Auxiliary Calculations

C.1 Topological Projectors

Let the network edges be sorted in such a way that the (reduced) incidence matrix A has the form

$$A = \begin{bmatrix} A_C & A_{\overline{R}} & A_{\overline{L}} & A_{\overline{V}} & A_I \end{bmatrix},$$

where the index stands for capacitive, (extended) resistive, (extended) inductive, (extended) voltage source and current source edges, respectively.

In order to describe different circuit configuration in more detail we will introduce some useful projectors.

We denote by

$$Q_C, \ Q_{C-\overline{V}}, \ Q_{\overline{V}-C}, \ Q_{\overline{R}-C\overline{V}} \text{ and } Q_{C\overline{R}\overline{V}}$$

projectors onto

$$\ker A_C^\top, \ \ker Q_C^\top A_{\overline{V}}, \ \ker A_{\overline{V}}^\top Q_C, \ \ker A_{\overline{R}}^\top Q_C Q_{\overline{V}-C} \text{ and } \ker \begin{bmatrix} A_C & A_{\overline{R}} & A_{\overline{V}} \end{bmatrix}^\top,$$

respectively. All complementary projectors will be denoted by $P = I - Q$ with corresponding subindex.

In the following we show that $Q_{C\overline{R}\overline{V}} = Q_C Q_{\overline{V}-C} Q_{\overline{R}-C\overline{V}}$ is a valid construction. That special construction goes back to [ET00], here it is slightly extended them to an enlarge class of network edges.

At first we introduce some notation and results concerning special cutsets and loops to motivate the projectors above.

Definition C.1 ($\overline{L}I$-cutset). A cutset is called $\overline{L}I$-cutset if and only if the cutset contains (extended) inductors and current sources only.

Lemma C.2 ($\overline{L}I$-cutsets). Let a connected circuit be given. The circuit does not contain an $\overline{L}I$-cutset if and only if

(i) the matrix $\begin{bmatrix} A_C & A_{\overline{R}} & A_{\overline{V}} \end{bmatrix}$ has full row rank or

(ii) the projector $Q_{C\overline{R}\overline{V}}$ is equal to the zero matrix.

Proof. See Lemma 1.2 in [Tis04]. □

Definition C.3 ($C\overline{V}$-loop). A loop is called $C\overline{V}$-loop if and only if the loop contains capacitors and (extended) voltage sources only.

Lemma C.4 ($C\overline{V}$-loops). Let Q_C be a projector onto $\ker A_C^\top$. The circuit does not contain a $C\overline{V}$-loop with at least one (extended) voltage source if and only if

(i) the matrix $Q_C^\top A_{\overline{V}}$ has full column rank or

(ii) the projector $Q_{C-\overline{V}}$ is equal to the zero matrix.

Proof. See Lemma 1.3 in [Tis04]. $\qquad\qquad\qquad\qquad\qquad\qquad\qquad\qquad\qquad\qquad$ □

Next we construct the projector $Q_{C\overline{RV}}$. For that we need the following lemmata.

Lemma C.5. The relations

(i) $\operatorname{im} P_C \subset \ker P_{\overline{V}-C}$

(ii) $\operatorname{im} P_{\overline{V}-C} \subset \ker P_{\overline{R}-C\overline{V}}$

(iii) $\operatorname{im} P_C \subset \ker P_{\overline{R}-C\overline{V}}$

hold true.

Proof. Straightforward computations show the results.

(i) $x \in \ker Q_C \Rightarrow A_{\overline{V}}^\top Q_C x = 0 \Rightarrow x \in \operatorname{im} Q_{\overline{V}-C}$

(ii) $x \in \ker Q_{\overline{V}-C} \Rightarrow A_{\overline{R}}^\top Q_C Q_{\overline{V}-C} x = 0 \Rightarrow x \in \operatorname{im} Q_{\overline{R}-C\overline{V}}$

(iii) $x \in \ker Q_C \overset{(i)}{\Rightarrow} A_{\overline{R}}^\top Q_C Q_{\overline{V}-C} x = 0 \Rightarrow x \in \operatorname{im} Q_{\overline{R}-C\overline{V}}$

$\qquad\qquad\qquad\qquad\qquad\qquad\qquad\qquad\qquad\qquad\qquad\qquad\qquad\qquad\qquad$ □

Corollary C.6. The relations

(i) $P_{\overline{V}-C} P_C = 0 \Leftrightarrow Q_{\overline{V}-C} Q_C = Q_C + Q_{\overline{V}-C} - I$

(ii) $P_{\overline{R}-C\overline{V}} P_C = 0 \Leftrightarrow Q_{\overline{R}-C\overline{V}} Q_C = Q_C + Q_{\overline{R}-C\overline{V}} - I$

(iii) $P_{\overline{R}-C\overline{V}} P_{\overline{V}-C} = 0 \Leftrightarrow Q_{\overline{R}-C\overline{V}} Q_{\overline{V}-C} = Q_{\overline{V}-C} + Q_{\overline{R}-C\overline{V}} - I$

hold true. $\qquad\qquad\qquad\qquad\qquad\qquad\qquad\qquad\qquad\qquad\qquad\qquad\qquad$ □

Lemma C.7. $Q_C Q_{\overline{V}-C}$ is a projector.

Proof. Using Corollary C.6 we get

$$\left(Q_C Q_{\overline{V}-C}\right)^2 = Q_C \left(Q_C + Q_{\overline{V}-C} - I\right) Q_{\overline{V}-C} = Q_C Q_{\overline{V}-C}.$$

$\qquad\qquad\qquad\qquad\qquad\qquad\qquad\qquad\qquad\qquad\qquad\qquad\qquad\qquad\qquad$ □

Lemma C.8. $Q_C Q_{\overline{V}-C} Q_{\overline{R}-C\overline{V}}$ is a projector.

Proof. Using Corollary C.6 and Lemma C.7 we get

$$
\begin{aligned}
\left(Q_C Q_{V-C} Q_{\overline{R}-C\overline{V}}\right)^2 &= Q_C Q_{\overline{V}-C}\left(Q_C + Q_{\overline{R}-C\overline{V}} - I\right) Q_{\overline{V}-C} Q_{\overline{R}-C\overline{V}} \\
&= Q_C Q_{\overline{V}-C} Q_{\overline{R}-C\overline{V}} Q_{\overline{V}-C} Q_{\overline{R}-C\overline{V}} \\
&= Q_C Q_{\overline{V}-C}\left(Q_{\overline{V}-C} + Q_{\overline{R}-C\overline{V}} - I\right) Q_{\overline{R}-C\overline{V}} \\
&= Q_C Q_{\overline{V}-C} Q_{\overline{R}-C\overline{V}}.
\end{aligned}
$$

$\square$

Lemma C.9. The relations $\operatorname{im} Q_C Q_{\overline{V}-C} Q_{\overline{R}-C\overline{V}} \subset \operatorname{im} Q_C$ hold true. $\square$

Lemma C.10. The matrix $Q_C Q_{\overline{V}-C} Q_{\overline{R}-C\overline{V}}$ is a projector onto $\ker\begin{bmatrix} A_C & A_{\overline{R}} & A_{\overline{V}} \end{bmatrix}^T$.

Proof. We have to show two inclusions.

($\subseteq$) Let be $x \in \operatorname{im} Q_C Q_{\overline{V}-C} Q_{\overline{R}-C\overline{V}}$. Then $x \in \operatorname{im} Q_C = \ker A_C^T$. We obtain $x \in \ker A_{\overline{R}}^T$ since $\operatorname{im} Q_{\overline{R}-C\overline{V}} = \ker A_{\overline{R}}^T Q_C Q_{\overline{V}-C}$. Due to $\operatorname{im} Q_{\overline{V}-C} = \ker A_{\overline{V}}^T Q_C$ we get $x \in \ker A_{\overline{V}}^T$ and we conclude $\operatorname{im} Q_C Q_{\overline{V}-C} Q_{\overline{R}-C\overline{V}} \subset \ker\begin{bmatrix} A_C & A_{\overline{R}} & A_{\overline{V}} \end{bmatrix}^T$.

($\supseteq$) Let be $x \in \ker\begin{bmatrix} A_C & A_{\overline{R}} & A_{\overline{V}} \end{bmatrix}^T$. Then we get $x \in \operatorname{im} Q_C$ and also $x \in \ker A_{\overline{V}}^T Q_C$. Consequently, we gain $x \in \operatorname{im} Q_{\overline{V}-C}$ and thus $x \in \operatorname{im} Q_C Q_{\overline{V}-C}$. From $x \in \ker A_{\overline{R}}^T$ we obtain $x \in \ker A_{\overline{R}}^T Q_C Q_{\overline{V}-C}$ and hence $x \in \operatorname{im} Q_{\overline{R}-C\overline{V}}$. Accordingly, we achieve $x \in \operatorname{im} Q_C Q_{\overline{V}-C} Q_{\overline{R}-C\overline{V}}$ and therefore $\ker\begin{bmatrix} A_C & A_{\overline{R}} & A_{\overline{V}} \end{bmatrix}^T \subset \operatorname{im} Q_C Q_{\overline{V}-C} Q_{\overline{R}-C\overline{V}}$.

$\square$

Lemma C.11. The relation $\ker Q_C \subset \ker Q_{C\overline{R}\overline{V}}$ hold true.

Proof. We use Lemma C.5. Let be $x \in \ker Q_C$. Then $x \in \operatorname{im} Q_{\overline{V}-C}$ and we achieve $x \in \ker Q_C Q_{\overline{V}-C}$. Therefore $x \in \operatorname{im} Q_{\overline{R}-C\overline{V}}$ and thus $x \in \ker Q_{C\overline{R}\overline{V}}$. $\square$

Corollary C.12. The matrix

$$
Q_{C\overline{R}\overline{V}} = Q_C Q_{\overline{V}-C} Q_{\overline{R}-C\overline{V}}
$$

is a projector onto $\ker\begin{bmatrix} A_C & A_{\overline{R}} & A_{\overline{V}} \end{bmatrix}^T$ and the relation $Q_{C\overline{R}\overline{V}} Q_C = Q_{C\overline{R}\overline{V}}$ holds true. $\square$

C.2 Electric Network

Lemma C.13. Let Assumption 4.4 and 4.9 be fulfilled. For the DAE (4.8) we get

$$
W_0(y, t) = \begin{bmatrix} Q_C^T & -Q_C^T A_M M\left(q_M, t\right)^{-1} & 0 & 0 \\ 0 & 0 & 0 & 0 \\ 0 & 0 & 0 & 0 \\ 0 & 0 & 0 & I \end{bmatrix}.
$$

Proof. We compute a projector along $\operatorname{im} G_0(y, t)$. In order to determine such a projector we calculate a projector onto $\ker G_0(y, t)^\top$, see Remark A.8, with

$$
G_0(y, t)^\top = \begin{bmatrix} A_C C \left(A_C^\top, t\right)^\top A_C^\top & 0 & 0 & 0 \\ A_M^\top & M\left(q_M, t\right)^\top & 0 & 0 \\ 0 & 0 & L\left(j_L, t\right)^\top & 0 \\ 0 & 0 & 0 & 0 \end{bmatrix},
$$

see (4.9). Let be $z \in \ker G_0(y, t)^\top$. This it true if and only if

$$
z_e \in \operatorname{im} Q_C
$$
$$
-M\left(q_M, t\right)^{-\top} A_M^\top z_{e,t} = z_{q_M}
$$
$$
z_{j_L} = 0
$$

hold true, using Assumption 4.4, 4.9 and Lemma A.3. We can choose a projector onto $\ker G_0(y, t)^\top$ and along $\operatorname{im} G_0(y, t)$ by

$$
W_0(y, t)^\top = \begin{bmatrix} Q_C & 0 & 0 & 0 \\ -M\left(q_M, t\right)^{-\top} A_M^\top Q_C & 0 & 0 & 0 \\ 0 & 0 & 0 & 0 \\ 0 & 0 & 0 & I \end{bmatrix}
$$

and

$$
W_0(y, t) = \begin{bmatrix} Q_C^\top & -Q_C^\top A_M M\left(q_M, t\right)^{-1} & 0 & 0 \\ 0 & 0 & 0 & 0 \\ 0 & 0 & 0 & 0 \\ 0 & 0 & 0 & I \end{bmatrix},
$$

respectively. □

Lemma C.14. Let Assumption 4.4 and 4.9 be fulfilled. For the DAE (4.8) we get

$$
W_1 = \begin{bmatrix} Q_{CRMV}^\top & 0 & 0 & 0 \\ 0 & 0 & 0 & 0 \\ 0 & 0 & 0 & 0 \\ 0 & 0 & 0 & Q_{C-V}^\top \end{bmatrix}.
$$

Proof. We compute a projector along $\operatorname{im} G_1(y, t)$. On this we determine a projector onto $\ker G_1(y, t)^\top$, see Remark A.8. Hereby we investigate

$$
G_1(y, t)^\top = \begin{bmatrix} \overline{G}(e, t)^\top & -Q_C^\top A_M & -Q_C^\top A_L & Q_C^\top A_V \\ A_M^\top & M\left(q_M, t\right)^\top & 0 & 0 \\ 0 & 0 & L\left(j_L, t\right)^\top & 0 \\ A_V^\top & 0 & 0 & 0 \end{bmatrix},
$$

with $\overline{G}(e,t) = A_C C \left(A_C^T e, t\right)^T A_C^T + A_R G \left(A_R^T e, t\right)^T A_R^T Q_C$, see (4.13).

Let be $z \in \ker G_1(y,t)^T$. That is true if and only if

$$\left(A_C C \left(A_C^T e, t\right)^T A_C^T + Q_C^T A_R G \left(A_R^T e, t\right)^T A_R^T\right) z_e - Q_C^T A_M z_{q_M} + Q_C^T A_V z_{j_V} = 0 \qquad (C.1)$$

$$-M(q_M, t)^{-T} A_M^T z_e = z_{q_M} \qquad (C.2)$$

$$z_{j_L} = 0$$

$$A_V^T z_e = 0 \qquad (C.3)$$

hold true, taking Assumption 4.4 and 4.9 into account. Left-multiplication of (C.1) by Q_C^T yields

$$Q_C^T A_R G \left(A_R^T e, t\right)^T A_R^T z_e - Q_C^T A_M z_{q_M} + Q_C^T A_V z_{j_V} = 0$$

and subtraction of (C.1) leads to $z_e \in \operatorname{im} Q_C$ by applying Lemma A.3. Using $z_e \in \operatorname{im} Q_C$ and (C.2) we can rewrite (C.1) as

$$Q_C^T A_R G \left(A_R^T e, t\right)^T A_R^T Q_C z_e + Q_C^T A_M M(q_M, t)^{-T} A_M^T Q_C z_e + Q_C^T A_V z_{j_V} = 0.$$

Left-multiplying that by z_e^T leads to $z_e \in \ker \begin{bmatrix} A_R & A_M \end{bmatrix}^T$ by taking (C.3) into account. Together with (C.3) we attain $z_e \in \operatorname{im} Q_{CRMV}$. From (C.2) we obtain $z_{q_M} = 0$ and (C.1) yields

$$Q_C^T A_V z_{j_V} = 0 \text{ and } z_{j_V} \in \operatorname{im} Q_{C-V},$$

that is, we can choose a projector onto $\ker G_1(y,t)^T$ and along $\operatorname{im} G_1(y,t)$ by

$$W_1^T = \begin{bmatrix} Q_{CRMV} & 0 & 0 & 0 \\ 0 & 0 & 0 & 0 \\ 0 & 0 & 0 & 0 \\ 0 & 0 & 0 & Q_{C-V} \end{bmatrix} \text{ and } W_1 = \begin{bmatrix} Q_{CRMV}^T & 0 & 0 & 0 \\ 0 & 0 & 0 & 0 \\ 0 & 0 & 0 & 0 \\ 0 & 0 & 0 & Q_{C-V}^T \end{bmatrix},$$

respectively. $\qquad \square$

Lemma C.15. Let Assumption 4.4 and 4.9 be fulfilled. For the DAE (4.8) we get

$$\mathcal{N}_1(y,t) = \{z \in \mathbb{R}^n | Q_C z_e \in \operatorname{im} Q_{CRMV}, z_{j_V} \in \operatorname{im} Q_{C-V}, z_{q_M} = 0,$$
$$P_C z_e = -H_C \left(A_C^T e, t\right)^{-1} A_V Q_{C-V} z_{j_V}, L(j_L, t)^{-1} A_L^T Q_C z_e = z_{j_L}\}.$$

Proof. Let be $z \in \ker G_1(y,t)$. This is holds if and only if

$$A_C C \left(A_C^T e, t\right) A_C^T + A_R G \left(A_R^T e, t\right) A_R^T Q_C z_e + A_M z_{q_M} + A_V z_{j_V} = 0, \qquad (C.4)$$

$$M(q_M, t)^{-1} A_M^T Q_C z_e = z_{q_M} \qquad (C.5)$$

$$L\left(j_L, t\right)^{-1} A_L^T Q_C z_e = z_{j_L}$$
$$A_V^T Q_C z_e = 0, \qquad (C.6)$$

are valid, see (4.13) and using Assumption 4.4 and 4.9. Left-multiplication of (C.4) by $z_e^T Q_C^T$ and inserting (C.5) give rise to

$$z_e^T Q_C^T A_R G\left(A_R^T e, t\right) A_R^T Q_C z_e + z_e^T Q_C^T A_M M\left(q_M, t\right)^{-1} A_M^T Q_C z_e + z_e^T Q_C^T A_V z_{jv} = 0$$

Utilizing (C.6) leads to $Q_C z_e \in \ker \begin{bmatrix} A_R & A_M \end{bmatrix}^T$ by using Lemma A.3. We attain $z_{q_M} = 0$ and together with (C.6) to $Q_C z_e \in \operatorname{im} Q_{CRMV}$. Left-multiplication of (C.4) by Q_C^T results in

$$Q_C^T A_V z_{jv} = 0 \text{ and } z_{jv} \in \operatorname{im} Q_{C-V}.$$

Moreover we can reduce (C.4) to

$$A_C C\left(A_C^T e, t\right) A_C^T z_e + A_V z_{jv} = 0$$

which can be rewritten as

$$H_C\left(A_C^T e, t\right) P_C z_e + A_V z_{jv} = 0,$$

where $H_C\left(A_C^T e, t\right) = A_C C\left(A_C^T e, t\right) A_C^T + Q_C^T Q_C$ is positive definite, see Lemma A.10. Hence we deduce that

$$Q_C z_e \in \operatorname{im} Q_{CRMV}$$
$$z_{jv} \in \operatorname{im} Q_{C-V}$$
$$z_{q_M} = 0$$
$$P_C z_e = -H_C\left(A_C^T e, t\right)^{-1} A_V Q_{C-V} z_{jv}$$
$$L\left(j_L, t\right)^{-1} A_L^T Q_C z_e = z_{j_L}$$

hold true and yields the representation

$$\mathcal{N}_1\left(y, t\right) = \{z \in \mathbb{R}^n | Q_C z_e \in \operatorname{im} Q_{CRMV}, z_{jv} \in \operatorname{im} Q_{C-V}, z_{q_M} = 0,$$
$$P_C z_e = -H_C\left(A_C^T e, t\right)^{-1} A_V Q_{C-V} z_{jv}, L\left(j_L, t\right)^{-1} A_L^T Q_C z_e = z_{j_L}\}.$$

$\square$

C.3 Field/Circuit System using Coulomb Gauge

Lemma C.16. Assume Assumption 4.4, 3.31 and Property 3.32 to be fulfilled. For the DAE (5.2) we get

$$W_0 = \begin{bmatrix} Q_C^T & 0 & 0 & 0 & 0 & 0 & 0 \\ 0 & 0 & 0 & 0 & 0 & 0 & 0 \\ 0 & 0 & I & 0 & 0 & 0 & 0 \\ 0 & 0 & 0 & I & 0 & 0 & 0 \\ 0 & 0 & 0 & 0 & I & 0 & 0 \\ 0 & 0 & 0 & 0 & 0 & 0 & 0 \\ 0 & 0 & 0 & 0 & 0 & 0 & 0 \end{bmatrix}.$$

Proof. We compute a projector along $\operatorname{im} G_0(y, t)$. In order to determine such a projector we calculate a projector onto $\ker G_0(y, t)^\top$, see Remark A.8, with

$$
G_0(y, t)^\top =
\begin{bmatrix}
A_C C \left(A_C^\top e, t\right)^\top A_C^\top & 0 & 0 & 0 & 0 & -A_E \Lambda_u^\top M_\varepsilon^u & 0 \\
0 & L\left(j_L, t\right)^\top & 0 & 0 & 0 & 0 & 0 \\
0 & 0 & 0 & 0 & 0 & 0 & 0 \\
0 & 0 & 0 & 0 & 0 & 0 & 0 \\
0 & 0 & 0 & 0 & 0 & -\widetilde{S}_u M_\varepsilon^u & 0 \\
0 & 0 & 0 & 0 & 0 & 0 & I \\
0 & 0 & 0 & 0 & 0 & M_\varepsilon^u & 0
\end{bmatrix},
$$

see (5.3), using Assumption 3.31 and Property 3.32. Let be $z \in \ker G_0(y, t)^\top$. This is valid if and only if

$$
z_e \in \operatorname{im} Q_C
$$
$$
z_{j_L} = 0
$$
$$
z_{\widehat{a}_u} = 0
$$
$$
z_{\widehat{\pi}_u} = 0
$$

hold true, using Assumption 4.4 and Lemma A.3. We can choose a projector onto $\ker G_0(y, t)^\top$ and along $\operatorname{im} G_0(y, t)$ by

$$
W_0^\top =
\begin{bmatrix}
Q_C & 0 & 0 & 0 & 0 & 0 & 0 \\
0 & 0 & 0 & 0 & 0 & 0 & 0 \\
0 & 0 & I & 0 & 0 & 0 & 0 \\
0 & 0 & 0 & I & 0 & 0 & 0 \\
0 & 0 & 0 & 0 & I & 0 & 0 \\
0 & 0 & 0 & 0 & 0 & 0 & 0 \\
0 & 0 & 0 & 0 & 0 & 0 & 0
\end{bmatrix}
\quad \text{and} \quad
W_0 =
\begin{bmatrix}
Q_C^\top & 0 & 0 & 0 & 0 & 0 & 0 \\
0 & 0 & 0 & 0 & 0 & 0 & 0 \\
0 & 0 & I & 0 & 0 & 0 & 0 \\
0 & 0 & 0 & I & 0 & 0 & 0 \\
0 & 0 & 0 & 0 & I & 0 & 0 \\
0 & 0 & 0 & 0 & 0 & 0 & 0 \\
0 & 0 & 0 & 0 & 0 & 0 & 0
\end{bmatrix},
$$

respectively. $\qquad\square$

Lemma C.17. Let Assumption 4.4, 3.31 and Property 3.32 be true. For the DAE (5.2) we get

$$
W_1 =
\begin{bmatrix}
Q_{CRV}^\top & 0 & 0 & -Q_{CRV}^\top A_E & 0 & 0 & 0 \\
0 & 0 & 0 & 0 & 0 & 0 & 0 \\
0 & 0 & Q_{C-V}^\top & 0 & 0 & 0 & 0 \\
0 & 0 & 0 & 0 & 0 & 0 & 0 \\
0 & 0 & 0 & 0 & I & 0 & 0 \\
0 & 0 & 0 & 0 & 0 & 0 & 0 \\
0 & 0 & 0 & 0 & 0 & 0 & 0
\end{bmatrix}.
$$

Proof. We compute a projector along $\operatorname{im} G_1(y,t)$. On this we determine a projector onto $\ker G_1(y,t)^\top$, see Remark A.8. Hereby we investigate

$$
G_1(y,t)^\top = \begin{bmatrix}
\overline{G}(e,t)^\top & -Q_C^\top A_L & Q_C^\top A_V & 0 & 0 & -A_E \Lambda_u^\top M_\varepsilon^u & -Q_C^\top A_E \Lambda_u^\top \\
0 & L(j_L,t)^\top & 0 & 0 & 0 & 0 & 0 \\
A_V^\top & 0 & 0 & 0 & 0 & 0 & 0 \\
A_E^\top & 0 & 0 & I & 0 & 0 & 0 \\
0 & 0 & 0 & 0 & 0 & -\widetilde{S}_u M_\varepsilon^u & -\widetilde{S}_u \\
0 & 0 & 0 & 0 & 0 & 0 & I \\
0 & 0 & 0 & 0 & 0 & M_\varepsilon^u & 0
\end{bmatrix},
$$

where $\overline{G}(e,t) = A_C C\left(A_C^\top e,t\right) A_C^\top + Q_C^\top A_R G\left(A_R^\top e,t\right) A_R^\top$, see (5.7), using Assumption 3.31 and Property 3.32. Let be $z \in \ker G_1(y,t)^\top$. This is valid if and only if

$$
\left(A_C C\left(A_C^\top e,t\right)^\top A_C^\top + Q_C^\top A_R G\left(A_R^\top e,t\right)^\top A_R^\top\right) z_e + Q_C^\top A_V z_{j_V} = 0 \tag{C.7}
$$

$$
z_{j_L} = 0
$$

$$
A_V^\top z_e = 0 \tag{C.8}
$$

$$
z_{j_E} = -A_E^\top z_e
$$

$$
z_{\widehat{a}_u} = 0
$$

$$
z_{\widehat{\pi}_u} = 0
$$

hold true, taking Assumption 4.4 into account. Left-multiplication of (C.7) by $Q_C^\top$ yields

$$
Q_C^\top A_R G\left(A_R^\top e,t\right)^\top A_R^\top z_e + Q_C^\top A_V z_{j_V} = 0 \tag{C.9}
$$

and subtraction from (C.7) leads to

$$
A_C C\left(A_C^\top e,t\right)^\top A_C^\top z_e = 0.
$$

Hence we obtain $z_e \in \operatorname{im} Q_C$ due to Lemma A.3. Left-multiply (C.9) by $z_e^\top$ and taking (C.8) into account leads to $z_e \in \operatorname{im} Q_{CRV}$. From (C.9) follows that $z_{j_V} \in \operatorname{im} Q_{C-V}$. Hence we deduce

$$
z_e \in \operatorname{im} Q_{CRV}
$$

$$
z_{j_L} = 0
$$

$$
z_{j_V} \in \operatorname{im} Q_{C-V}
$$

$$
z_{j_E} = -A_E^\top Q_{CRV} z_e
$$

$$
z_{\widehat{a}_u} = 0
$$

$$
z_{\widehat{\pi}_u} = 0
$$

and we can choose a projector onto $\ker G_1 (y, t)^\top$ and along $\operatorname{im} G_1 (y, t)$ by

$$
W_1^\top = \begin{bmatrix}
Q_{CRV} & 0 & 0 & 0 & 0 & 0 & 0 \\
0 & 0 & 0 & 0 & 0 & 0 & 0 \\
0 & 0 & Q_{C-V} & 0 & 0 & 0 & 0 \\
-A_E^\top Q_{CRV} & 0 & 0 & 0 & 0 & 0 & 0 \\
0 & 0 & 0 & 0 & I & 0 & 0 \\
0 & 0 & 0 & 0 & 0 & 0 & 0 \\
0 & 0 & 0 & 0 & 0 & 0 & 0
\end{bmatrix}
$$

and

$$
W_1 = \begin{bmatrix}
Q_{CRV}^\top & 0 & 0 & -Q_{CRV}^\top A_E & 0 & 0 & 0 \\
0 & 0 & 0 & 0 & 0 & 0 & 0 \\
0 & 0 & Q_{C-V}^\top & 0 & 0 & 0 & 0 \\
0 & 0 & 0 & 0 & 0 & 0 & 0 \\
0 & 0 & 0 & 0 & I & 0 & 0 \\
0 & 0 & 0 & 0 & 0 & 0 & 0 \\
0 & 0 & 0 & 0 & 0 & 0 & 0
\end{bmatrix},
$$

respectively. $\qquad\square$

Lemma C.18. Let Assumption 4.4, 3.31 and Property 3.32 be true. For the DAE (5.2) we get

$$
\mathcal{N}_1 (y, t) = \big\{ z \in \mathbb{R}^n \mid Q_C z_e \in \operatorname{im} Q_{CRV}, \ z_{jV} \in \operatorname{im} Q_{C-V},
$$
$$
L (j_L, t)^{-1} A_L^\top Q_C z_e = z_{jL}, \ P_C z_e = -H_C \left(A_C^\top e, t \right)^{-1} A_V Q_{C-V} z_{jV}, \ z_{jE} = 0,
$$
$$
\Lambda_u A_E^\top z_e - G_u z_{\phi_u} = z_{\widehat{\pi}_u}, \ \Lambda_u A_E^\top Q_C z_e - G_u z_{\phi_u} = z_{\widehat{a}_u} \big\}.
$$

Proof. Let be $z \in \ker G_1 (y, t)$. This is valid if and only if

$$
\left(A_C C \left(A_C^\top e, t \right) A_C^\top + A_R G \left(A_R^\top e, t \right) A_R^\top Q_C \right) z_e + A_V z_{jV} = 0 \tag{C.10}
$$
$$
L (j_L, t)^{-1} A_L^\top Q_C z_e = z_{jL}
$$
$$
A_V^\top Q_C z_e = 0 \tag{C.11}
$$
$$
z_{jE} = 0
$$
$$
\Lambda_u A_E^\top z_e - G_u z_{\phi_u} = z_{\widehat{\pi}_u}
$$
$$
\Lambda_u A_E^\top Q_C z_e - G_u z_{\phi_u} = z_{\widehat{a}_u}
$$

hold true, see (5.3), using Assumption 4.4, 3.31 and Property 3.32. Left-multiplication of (C.10) by $z_e^\top Q_C^\top$ and taking advantage of (C.11) results in $Q_C z_e \in \ker A_R^\top$ due to Lemma A.3. From (C.11) we attain $Q_C z_e \in \operatorname{im} Q_{CRV}$. With this (C.10) reduces to

$$
A_C C \left(A_C^\top e, t \right) A_C^\top z_e + A_V z_{jV} = 0 \tag{C.12}
$$

which can be rewritten as

$$H_C\left(A_C^\top e, t\right) P_C z_e + A_V z_{j_V} = 0,$$

where $H_C\left(A_C^\top e, t\right) = A_C C\left(A_C^\top e, t\right) A_C^\top + Q_C^\top Q_C$ is positive definite, see Lemma A.10. Left-multiplication of (C.12) by $Q_C^\top$ leads to $z_{j_V} \in \operatorname{im} Q_{C-V}$. Hence we deduce that

$$Q_C z_e \in \operatorname{im} Q_{CRV}$$
$$z_{j_V} \in \operatorname{im} Q_{C-V}$$
$$z_{j_E} = 0$$
$$P_C z_e = -H_C\left(A_C^\top e, t\right)^{-1} A_V Q_{C-V} z_{j_V}$$
$$L\left(j_L, t\right)^{-1} A_L^\top Q_C z_e = z_{j_L}$$
$$\Lambda_u A_E^\top z_e - G_u z_{\phi_u} = z_{\widehat{\pi}_u}$$
$$\Lambda_u A_E^\top Q_C z_e - G_u z_{\phi_u} = z_{\widehat{a}_u}$$

hold true and we obtain

$$\mathcal{N}_1\left(y, t\right) = \Big\{ z \in \mathbb{R}^n \mid Q_C z_e \in \operatorname{im} Q_{CRV}, \ z_{j_V} \in \operatorname{im} Q_{C-V},$$
$$L\left(j_L, t\right)^{-1} A_L^\top Q_C z_e = z_{j_L}, \ P_C z_e = -H_C\left(A_C^\top e, t\right)^{-1} A_V Q_{C-V} z_{j_V}, \ z_{j_E} = 0,$$
$$\Lambda_u A_E^\top z_e - G_u z_{\phi_u} = z_{\widehat{\pi}_u}, \ \Lambda_u A_E^\top Q_C z_e - G_u z_{\phi_u} = z_{\widehat{a}_u} \Big\}.$$

$\square$

C.4 Field/Circuit System using Lorenz Gauge

Lemma C.19. Let Assumption 4.4, 3.31 and Property 3.32 be true. For the DAE (5.18) we get

$$W_0 = \begin{bmatrix} Q_C^\top & 0 & 0 & 0 & 0 & 0 & 0 \\ 0 & 0 & 0 & 0 & 0 & 0 & 0 \\ 0 & 0 & I & 0 & 0 & 0 & 0 \\ 0 & 0 & 0 & I & 0 & 0 & 0 \\ 0 & 0 & 0 & 0 & 0 & 0 & 0 \\ 0 & 0 & 0 & 0 & 0 & 0 & 0 \\ 0 & 0 & 0 & 0 & 0 & 0 & 0 \end{bmatrix}.$$

Proof. We compute a projector onto $\ker G_0\left(y, t\right)$. On this we determine a projector along $\ker G_0\left(y, t\right)^\top$, see Remark A.8. Hereby we investigate

$$G_0\left(y, t\right)^\top = \begin{bmatrix} A_C C\left(A_C^\top e, t\right)^\top A_C^\top & 0 & 0 & 0 & 0 & -A_E \Lambda_u^\top M_\varepsilon^u & 0 \\ 0 & L\left(j_L, t\right)^\top & 0 & 0 & 0 & 0 & 0 \\ 0 & 0 & 0 & 0 & 0 & 0 & 0 \\ 0 & 0 & 0 & 0 & 0 & 0 & 0 \\ 0 & 0 & 0 & 0 & \widetilde{S}_u M_\varepsilon^u G_u & -\widetilde{S}_u M_\varepsilon^u & 0 \\ 0 & 0 & 0 & 0 & 0 & 0 & I \\ 0 & 0 & 0 & 0 & 0 & M_\varepsilon^u & 0 \end{bmatrix},$$

see (5.19), using Assumption 3.31 and Property 3.32. Let be $z \in \ker G_0\left(y, t\right)^\top$. This is valid if and only if

$$z_e \in \operatorname{im} Q_C$$
$$z_{j_L} = 0$$
$$z_{\widehat{a}_u} = 0$$
$$z_{\widehat{\pi}_u} = 0$$
$$z_{\phi_u} = 0$$

hold true, due to Lemma A.3 and Assumption 4.4. Hence we can choose a projector onto $\ker G_0\left(y, t\right)^\top$ and along $\operatorname{im} G_0\left(y, t\right)$ by

$$W_0^\top = \begin{bmatrix} Q_C & 0 & 0 & 0 & 0 & 0 & 0 \\ 0 & 0 & 0 & 0 & 0 & 0 & 0 \\ 0 & 0 & I & 0 & 0 & 0 & 0 \\ 0 & 0 & 0 & I & 0 & 0 & 0 \\ 0 & 0 & 0 & 0 & 0 & 0 & 0 \\ 0 & 0 & 0 & 0 & 0 & 0 & 0 \\ 0 & 0 & 0 & 0 & 0 & 0 & 0 \end{bmatrix} \text{ and } W_0 = \begin{bmatrix} Q_C^\top & 0 & 0 & 0 & 0 & 0 & 0 \\ 0 & 0 & 0 & 0 & 0 & 0 & 0 \\ 0 & 0 & I & 0 & 0 & 0 & 0 \\ 0 & 0 & 0 & I & 0 & 0 & 0 \\ 0 & 0 & 0 & 0 & 0 & 0 & 0 \\ 0 & 0 & 0 & 0 & 0 & 0 & 0 \\ 0 & 0 & 0 & 0 & 0 & 0 & 0 \end{bmatrix}.$$

$\square$

Lemma C.20. Assume Assumption 4.4, 3.31 and Property 3.32 to be fulfilled. For the DAE (5.18) we get

$$W_1 = \begin{bmatrix} Q_{CRV}^\top & 0 & 0 & -Q_{CRV}^\top A_E & 0 & 0 & 0 \\ 0 & 0 & 0 & 0 & 0 & 0 & 0 \\ 0 & 0 & Q_{C-V}^\top & 0 & 0 & 0 & 0 \\ 0 & 0 & 0 & 0 & 0 & 0 & 0 \\ 0 & 0 & 0 & 0 & 0 & 0 & 0 \\ 0 & 0 & 0 & 0 & 0 & 0 & 0 \\ 0 & 0 & 0 & 0 & 0 & 0 & 0 \end{bmatrix}.$$

Proof. We compute a projector along $\operatorname{im} G_1\left(y, t\right)$. On this we determine a projector $W_1\left(y, t\right)^\top$ along $\ker G_1\left(y, t\right)^\top$, see Remark A.8, where

$$G_1\left(y, t\right)^\top = \begin{bmatrix} \overline{G}\left(e, t\right)^\top & -Q_C^\top A_L & Q_C^\top A_V & 0 & 0 & -A_E \Lambda_u^\top M_\varepsilon^u & -Q_C^\top A_E \Lambda_u^\top \\ 0 & L\left(j_L, t\right)^\top & 0 & 0 & 0 & 0 & 0 \\ A_V^\top & 0 & 0 & 0 & 0 & 0 & 0 \\ A_E^\top & 0 & 0 & I & 0 & 0 & 0 \\ 0 & 0 & 0 & 0 & \widetilde{S}_u M_\varepsilon^u G_u & -\widetilde{S}_u M_\varepsilon^u & 0 \\ 0 & 0 & 0 & 0 & 0 & 0 & I \\ 0 & 0 & 0 & 0 & 0 & M_\varepsilon^u & 0 \end{bmatrix},$$

see (5.20), with $\overline{G}(e,t) = A_C C \left(A_C^\top e, t \right) A_C^\top + A_R G \left(A_R^\top e, t \right) A_R^\top Q_C$, using Assumption 3.31 and Property 3.32. Let be $z \in \ker G_1 (y,t)^\top$. This is valid if and only if

$$\left(A_C C \left(A_C^\top e, t \right)^\top A_C^\top + Q_C^\top A_R G \left(A_R^\top e, t \right)^\top A_R^\top \right) z_e + Q_C^\top A_V z_{j_V} = 0 \tag{C.13}$$

$$z_{j_L} = 0$$

$$A_V^\top z_e = 0 \tag{C.14}$$

$$z_{j_E} = -A_E^\top z_e$$

$$z_{\widehat{a}_u} = 0$$

$$z_{\widehat{\pi}_u} = 0$$

$$z_{\phi_u} = 0$$

are valid, taking Assumption 4.4 into account. Left-multiplying (C.13) by $Q_C^\top$ yields

$$Q_C^\top A_R G \left(A_R^\top e, t \right)^\top A_R^\top z_e + Q_C^\top A_V z_{j_V} = 0 \tag{C.15}$$

and subtraction from (C.13) leads to

$$A_C C \left(A_C^\top e, t \right)^\top A_C^\top z_e = 0.$$

We obtain $z_e \in \operatorname{im} Q_C$ due to Lemma A.3. Left-multiplying (C.15) by $z_e^\top$ and taking (C.14) into account leads to $z_e \in \operatorname{im} Q_{CRV}$. Hence we deduce that

$$z_e \in \operatorname{im} Q_{CRV}$$

$$z_{j_L} = 0$$

$$z_{j_V} \in \operatorname{im} Q_{C-V}$$

$$z_{j_E} = -A_E^\top Q_{CRV} z_e$$

$$z_{\widehat{a}_u} = 0$$

$$z_{\widehat{\pi}_u} = 0$$

$$z_{\phi_u} = 0$$

hold true. We can choose a projector onto $\ker G_1 (y,t)^\top$ and along $\operatorname{im} G_1 (y,t)$ by

$$W_1^\top = \begin{bmatrix} Q_{CRV} & 0 & 0 & 0 & 0 & 0 & 0 \\ 0 & 0 & 0 & 0 & 0 & 0 & 0 \\ 0 & 0 & Q_{C-V} & 0 & 0 & 0 & 0 \\ -A_E^\top Q_{CRV} & 0 & 0 & 0 & 0 & 0 & 0 \\ 0 & 0 & 0 & 0 & 0 & 0 & 0 \\ 0 & 0 & 0 & 0 & 0 & 0 & 0 \\ 0 & 0 & 0 & 0 & 0 & 0 & 0 \end{bmatrix}$$

and

$$W_1 = \begin{bmatrix} Q_{CRV}^\top & 0 & 0 & -Q_{CRV}^\top A_E & 0 & 0 & 0 \\ 0 & 0 & 0 & 0 & 0 & 0 & 0 \\ 0 & 0 & Q_{C-V}^\top & 0 & 0 & 0 & 0 \\ 0 & 0 & 0 & 0 & 0 & 0 & 0 \\ 0 & 0 & 0 & 0 & 0 & 0 & 0 \\ 0 & 0 & 0 & 0 & 0 & 0 & 0 \\ 0 & 0 & 0 & 0 & 0 & 0 & 0 \end{bmatrix},$$

respectively. $\qquad\qquad\qquad\qquad\qquad\qquad\qquad\qquad\qquad\qquad\qquad\qquad\qquad\qquad\square$

Lemma C.21. Let Assumption 4.4, 3.31 and Property 3.32 be true. For the DAE (5.18) we get

$$\mathcal{N}_1(y,t) = \Big\{ z \in \mathbb{R}^n \mid Q_C z_e \in \operatorname{im} Q_{CRV}, \ z_{jv} \in \operatorname{im} Q_{C-V},$$
$$L(j_L,t)^{-1} A_L^\top Q_C z_e = z_{j_L}, \ P_C z_e = -H_C\left(A_C^\top e, t\right)^{-1} A_V Q_{C-V} z_{jv},$$
$$(z_{j_E}, z_{\phi_u}) = 0, \ \Lambda_u A_E^\top z_e = z_{\widehat{\pi}_u}, \ \Lambda_u A_E^\top Q_C z_e = z_{\widehat{a}_u} \Big\}.$$

Proof. Let be $z \in \ker G_1(y,t)$. This is valid if and only if

$$\left(A_C C\left(A_C^\top e, t\right) A_C^\top + A_R G\left(A_R^\top e, t\right) A_R^\top Q_C\right) z_e + A_V z_{jv} = 0 \qquad (C.16)$$
$$L(j_L,t)^{-1} A_L^\top Q_C z_e = z_{j_L}$$
$$A_V^\top Q_C z_e = 0 \qquad (C.17)$$
$$z_{j_E} = 0$$
$$\Lambda_u A_E^\top z_e = z_{\widehat{\pi}_u}$$
$$z_{\phi_u} = 0$$
$$\Lambda_u A_E^\top Q_C z_e = z_{\widehat{a}_u}$$

hold true, see (5.20), using Assumption 4.4, 3.31 and Property 3.32. Left-multiplying (C.16) by $z_e^\top Q_C^\top$ and taking (C.17) into account result in $Q_C z_e \in \ker A_R^\top$ due to Lemma A.3. We attain $Q_C z_e \in \operatorname{im} Q_{CRV}$ using (C.17). With this we reduce (C.16) to

$$A_C C\left(A_C^\top e, t\right) A_C^\top z_e + A_V z_{jv} = 0 \qquad (C.18)$$

which can be rewritten as

$$H_C\left(A_C^\top e, t\right) P_C z_e + A_V z_{jv} = 0,$$

where $H_C\left(A_C^\top e, t\right) = A_C C\left(A_C^\top e, t\right) A_C^\top + Q_C^\top Q_C$ is positive definite, see Lemma A.10. Left-multiplication of (C.18) by $Q_C^\top$ leads to $z_{jv} \in \operatorname{im} Q_{C-V}$. Hence we deduce that

$$Q_C z_e \in \operatorname{im} Q_{CRV}$$
$$z_{jv} \in \operatorname{im} Q_{C-V}$$

$$z_{j_E} = 0$$
$$z_{\phi_u} = 0$$
$$P_C z_e = -H_C \left(A_C^\top e, t \right)^{-1} A_V Q_{C-V} z_{j_V}$$
$$L \left(j_L, t \right)^{-1} A_L^\top Q_C z_e = z_{j_L}$$
$$\Lambda_u A_E^\top z_e = z_{\widehat{\pi}_u}$$
$$\Lambda_u A_E^\top Q_C z_e = z_{\widehat{a}_u}$$

are valid and in the end we achieve

$$\mathcal{N}_1 (y, t) = \Big\{ z \in \mathbb{R}^n \mid Q_C z_e \in \operatorname{im} Q_{CRV}, \ z_{j_V} \in \operatorname{im} Q_{C-V}, $$
$$L \left(j_L, t \right)^{-1} A_L^\top Q_C z_e = z_{j_L}, \ P_C z_e = -H_C \left(A_C^\top e, t \right)^{-1} A_V Q_{C-V} z_{j_V},$$
$$\left(z_{j_E}, z_{\phi_u} \right) = 0, \ \Lambda_u A_E^\top z_e = z_{\widehat{\pi}_u}, \ \Lambda_u A_E^\top Q_C z_e = z_{\widehat{a}_u} \Big\}.$$

$\square$

Notation

Abbreviation

ODE	ordinary differential equation
DAE	differential algebraic equation
MNA	modified nodal analysis
IVP	initial value problem
BDF	backward differentiation formulas
KCL	Kirchhoff's current law
KVL	Kirchhoff's voltage law
ME	Maxwell's equations
EM	electromagnetic
MQS	magnetoquasistatic
PEC	perfectly electric conducting
FIT	finite integration technique
MGE	Maxwell's grid equations
I-cutset	cutset of current sources
LI-cutset	cutset of inductors and current sources
LEI-cutset	cutset of inductors, EM devices and current sources
V-loop	loop of voltage sources
CV-loop	loop of capacitors and voltage sources

General

$\exists$	there exists
$\forall$	for all
$\mathbb{N}$	naturals
$\mathbb{Z}$	integer
$\mathbb{R}$	real numbers
$\mathbb{R}^n$	real n-dimensional vector space
$A \in \mathbb{R}^{n \times m}$ and $A \in \mathbb{Z}^{n \times m}$	matrix with n rows and m columns
I	identity matrix
$\mathcal{S}$	set
$|\mathcal{S}|$	number of elements in $\mathcal{S}$
$\dim \mathcal{S}$	dimension of a vector space $\mathcal{S}$
$\mathcal{S}^\perp$	orthogonal complement of $\mathcal{S} \subset \mathbb{R}^n$ with respect to the standard scalar product on $\mathbb{R}^n$
$\mathcal{I}$	interval
$\mathcal{D}$	domain of definition
$\{a_1, \ldots, a_n\}$	set consisting of the elements $a_1, \ldots, a_n$

$x \in \mathcal{M}$	x is an element of the set $\mathcal{M}$
$x \notin \mathcal{M}$	x is not an element of the set $\mathcal{M}$
$\mathcal{M} \subset \mathcal{N}$	$\mathcal{M}$ is contained in $\mathcal{N}$
$\mathcal{M} \not\subset \mathcal{N}$	$\mathcal{M}$ is not contained in $\mathcal{N}$
$\mathcal{M} \cup \mathcal{N}$	union of $\mathcal{M}$ and $\mathcal{N}$
$\mathcal{M} \cap \mathcal{N}$	intersection of $\mathcal{M}$ and $\mathcal{N}$
$\mathcal{M} \oplus \mathcal{N}$	direct sum of $\mathcal{M}$ and $\mathcal{N}$
$\mathcal{M} \times \mathcal{N}$	product set of $\mathcal{M}$ and $\mathcal{N}$
$f : \mathcal{M} \to \mathcal{N}$	function that maps from $\mathcal{M}$ into $\mathcal{N}$
$\frac{\partial}{\partial x} f$	partial derivative of f with respect to x
$\frac{d}{dt} f$	total derivative of f with respect to t
$\|\cdot\|$	function norm
$\mathcal{C}^k (\mathcal{M}, \mathcal{N})$	linear space of k-times ($k \geqslant 0$) continuously differentiable functions $f : \mathcal{M} \to \mathcal{N}$, $\mathcal{M} \subset \mathbb{R}^m$ and $\mathcal{N} \subset \mathbb{R}^n$ open
im A	image of the matrix A
ker A	kernel of the matrix A
rank A	rank of the matrix A
det A	determinant of the matrix A
$A^\top$	transpose of the matrix A
A^-	pseudoinverse of the matrix A
A^+	Moore-Penrose pseudoinverse of the matrix A
A^{-1}	inverse of the matrix A
$A^{-\top}$	transposed inverse of the matrix A
$\vec{n}_\perp$	normal vector
$\vec{n}_\shortparallel$	tangential vector

Matrix Chain and Subspaces

$\mathcal{C}_d^1 (\mathcal{I}, \mathcal{D})$	$= \{y \in \mathcal{C} (\mathcal{I}, \mathcal{D}) \mid d(y(\cdot), \cdot) \in \mathcal{C}^1 (\mathcal{I}, \mathbb{R}^m)\}$
$\mathcal{C}_D^1 (\mathcal{I}, \mathcal{D})$	$= \{y \in \mathcal{C} (\mathcal{I}, \mathcal{D}) \mid D(\cdot) y(\cdot) \in \mathcal{C}^1 (\mathcal{I}, \mathbb{R}^m)\}$
$D(y, t)$	$= \frac{\partial}{\partial y} d(y, t)$
$G_0 (y, t)$	$= A(y, t) D(y, t)$
$B_0 (z, y, t)$	$= \frac{\partial}{\partial y} [A(y, t) z + b(y, t)]$
$G_1 (z, y, t)$	$= G_0 (y, t) + B_0 (z, y, t) Q_0 (y, t)$
$\mathcal{N}_0 (y, t)$	$= \ker G_0 (y, t)$
$\mathcal{S}_0 (z, y, t)$	$= \{v \in \mathbb{R}^n \mid B_0 (z, y, t) v \in \operatorname{im} G_0 (y, t)\}$
$\mathcal{N}_1 (z, y, t)$	$= \ker G_1 (z, y, t)$
$\mathcal{S}_1 (z, y, t)$	$= \{v \in \mathbb{R}^n \mid B_0 (z, y, t) P_0 (y, t) v \in \operatorname{im} G_1 (y, t)\}$
$\mathcal{M}_0 (t)$ and $\mathcal{H}_1 (t)$	obvious and hidden constraint set
$\mathcal{M}_1 (t)$	index-2 constraint set

Projectors

$Q_0 (y, t)$	projector onto $\ker G_0 (y, t)$

$P_0(y,t)$	$= I - Q_0(y,t)$
$Q_1(z,y,t)$	projector onto $\ker G_1(z,y,t)$
$P_1(z,y,t)$	$= I - Q_1(z,y,t)$
$T(z,y,t)$	projector onto $\mathcal{N}_0(y,t) \cap \mathcal{S}_0(z,y,t)$
$U(z,y,t)$	$= I - T(z,y,t)$
$W_0(y,t)$	projector along $\operatorname{im} G_0(y,t)$
$W_1(z,y,t)$	projector along $\operatorname{im} G_1(z,y,t)$
$R(t)$	projector onto $\operatorname{im} D(y,t)$ and along $\ker A(y,t)$

Electric Network

$A_R, A_M, A_L, A_C, A_E, A_V, A_I$	incidence matrix of elements
e	node potential
q_M	charges through the memristors
j_M	currents through the memristors
j_L	currents through the inductors
j_V	currents through the voltage sources
j_E	currents through the electromagnetic devices
$v_s(t)$	given voltage sources
$i_s(t)$	given current sources
$q_C(u,t)$	charges of capacitors
$g_R(u,t)$	currents of resistors
$\phi_L(j,t)$	fluxes of inductors
$\phi_M(q,t)$	fluxes of memristors
$C(u,t)$	$= \frac{\partial}{\partial u} q_C(u,t)$
$G(u,t)$	$= \frac{\partial}{\partial u} g_R(u,t)$
$L(j,t)$	$= \frac{\partial}{\partial j} \phi_L(j,t)$
$M(q,t)$	$= \frac{\partial}{\partial q} \phi_M(q,t)$

Projectors for Electric Networks

Q_C	projector onto $\ker A_C^\top$
P_C	$= I - Q_C$
Q_{C-V}	projector onto $\ker Q_C^\top A_V$
P_{C-V}	$= I - Q_{C-V}$
Q_{V-C}	projector onto $\ker A_V^\top Q_C$
Q_{RM-CV}	projector onto $\ker \begin{bmatrix} A_R & A_M \end{bmatrix}^\top Q_C Q_{C-V}^\top$
Q_{CRMV}	projector onto $\ker \begin{bmatrix} A_C & A_V & A_R & A_M \end{bmatrix}^\top$
P_{CRMV}	$= I - Q_{CRMV}$
Q_{R-CV}	projector onto $\ker A_R^\top Q_C Q_{C-V}^\top$
Q_{CRV}	projector onto $\ker \begin{bmatrix} A_C & A_V & A_R \end{bmatrix}^\top$
P_{CRV}	$= I - Q_{CRV}$

Electromagnetic Field

$\vec{E}$	electric field

$\vec{H}$	magnetic field
$\vec{D}$	electric induction
$\vec{B}$	magnetic induction
ρ	distribution of charges
$\vec{J}_c$	conduction current density
$\vec{J}_d$	displacement current density
$\vec{J}_t$	total current density
ε	permittivity
ν	reluctivity
σ	conductivity
ζ	artificial material property
ξ	artificial material property
φ	scalar potential
$\vec{A}$	vector potential
$\vec{\Pi}$	auxiliary vector field

Discrete Electromagnetic Field

$\widehat{e}$, $\widehat{e}_u$	(reduced) discrete electric field strength
$\widehat{h}$, $\widehat{h}_u$	(reduced) discrete magnetic field strength
$\widehat{\widehat{d}}$, $\widehat{\widehat{d}}_u$	(reduced) discrete electric induction density
$\widehat{\widehat{b}}$, $\widehat{\widehat{b}}_u$	(reduced) discrete magnetic induction density
$\widehat{\widehat{q}}$, $\widehat{\widehat{q}}_u$	(reduced) discrete distribution of charges density
$\widehat{\widehat{j}}_c$, $\widehat{\widehat{j}}_{c,u}$	(reduced) discrete conduction current density
$\widehat{\widehat{j}}_t$, $\widehat{\widehat{j}}_{t,u}$	(reduced) discrete total current density
M_ε, M_ε^u	(reduced) discrete permittivity matrix
M_ν, M_ν^u	(reduced) discrete reluctivity matrix
M_σ, M_σ^u	(reduced) discrete conductivity matrix
M_ζ, M_ζ^u	(reduced) discrete artificial material property matrix
M_ξ, M_ξ^u	(reduced) discrete artificial material property matrix
ϕ, ϕ_u	(reduced) discrete scalar potential
$\widehat{a}$, $\widehat{a}_u$	(reduced) discrete vector potential
$\widehat{\pi}$, $\widehat{\pi}_u$	(reduced) discrete auxiliary vector
S, $\widetilde{S}$, S_u, $\widetilde{S}_u$	(reduced) discrete divergence operators
G, $\widetilde{G}$, G_u, $\widetilde{G}_u$	(reduced) discrete gradient operators
C, $\widetilde{C}$, C_u, $\widetilde{C}_u$	(reduced) discrete curl operators
Λ_u	excitation matrix

Vector Analysis

$\nabla\cdot$	divergence operator
∇	gradient operator
$\nabla\times$	curl operator
Δ	Laplace operator
∇^2	vector Laplace operator

Bibliography

[AASE+10] D. Abbott, S.F. Al-Sarawi, K. Eshraghian, A. Iqbal, O. Kavehei, and Y.S. Kim. The Fourth Element: Characteristics, Modelling, and Electromagnetic Theory of the Memristor. *Access*, 466(2120):28, 2010.

[AH01] U.M. Ascher and E. Haber. Fast finite volume simulation of 3D electromagnetic problems with highly discontinuous coefficients. *SIAM journal on Scientific Computing*, 22(6):1943–1961, 2001.

[AP98] U.M. Ascher and L.R. Petzold. *Computer methods for ordinary differential equations and differential-algebraic equations*, volume 61. SIAM, Philadelphia, 1998.

[Aré08] C. Arévalo. A note on numerically consistent initial values for high index differential-algebraic equations. *Electronic Transactions on Numerical Analysis*, 34:14–19, 2008.

[ASW95] M. Arnold, K. Strehmel, and R. Weiner. Errors in the numerical solution of nonlinear differential-algebraic systems of index 2. Technical report, Martin Luther Universität Halle-Wittenberg, Halle an der Saale, 1995.

[Bar04] A. Bartel. *Partial Differential-Algebraic Models in Chip Design - Thermal and Semiconductor Problems*. Fortschritt-Berichte VDI, Reihe 20. VDI Verlag, Düsseldorf, 2004. Dissertation.

[Bau08] S. Baumanns. Konsistente Initialisierung differential-algebraischer Gleichungen der gekoppelten Schaltungs- und Bauelementesimulation. Diplomarbeit, Universität zu Köln, 2008.

[BBB09a] D. Biolek, Z. Biolek, and V. Biolkova. SPICE model of memristor with nonlinear dopant drift. *Radioengineering*, 18(2):210–214, 2009.

[BBB09b] D. Biolek, Z. Biolek, and V. Biolkova. SPICE Modeling of Memristive, Memcapacitative and Meminductive Systems. *European Conference on Circuit Theory Design*, pages 249–252, 2009.

[BBBK10] D. Biolek, Z. Biolek, V. Biolkova, and Z. Kolka. Memristor modeling based on its constitutive relation. In *Proceedings of the European conference of systems, and European conference of circuits technology and devices, and*

European conference of communications, and European conference on Computer science, pages 261–264. World Scientific and Engineering Academy and Society, 2010.

[BBS11]   A. Bartel, S. Baumanns, and S. Schöps. Structural Analysis of Electrical Circuits Including Magnetoquasistatic Devices. *Applied Numerical Mathematics*, 61(12):1257–1270, 2011.

[BCDS11]   A. Bartel, M. Clemens, H. De Gersem, and S. Schöps. Decomposition and regularization of nonlinear anisotropic curl-curl DAEs. *COMPEL: The International journal for Computation and Mathematics in Electrical and Electronic Engineering*, 30(6):1701–1714, 2011.

[BCK+11]   S. Baumanns, T. Clees, B. Klaassen, M. Selva Soto, and C. Tischendorf. Fully Coupled Circuit and Device Simulation with Exploitation of Algebraic Multigrid Linear Solvers. In *Proceedings of the edaWorkshop 11*, pages 63–68. VDE Verlag, 2011.

[BCP96]   K.E. Brenan, S.L. Campbell, and L.R. Petzold. *Numerical solution of initial-value problems in differential-algebraic equations*, volume 14. SIAM, 1996.

[BCS12]   S. Baumanns, M. Clemens, and S. Schöps. Structural Aspects of Regularized Full Maxwell Electrodynamic Potential Formulations Using FIT. *Submitted to: 15th International IGTE Symposium on Numerical Field Calculation in Electrical Engineering*, 2012.

[BDD+92]   M. Bartsch, M. Dehler, M. Dohlus, F. Ebeling, P. Hahne, R. Klatt, F. Krawczyk, M. Marx, Z. Min, T. Propper, D. Schmitt, P. Schutt, B. Steffen, B. Wagner, T. Weiland, S.G. Wipf, and H. Wolter. Solution of Maxwell's Equations. *Computer Physics Communications*, 73(1-3):22–39, 1992.

[Ben06]   G. Benderskaya. *Numerical Methods for Transient Field-Circuit Coupled Simulations Based on the Finite Integration Technique and a Mixed Circuit Formulation*. Dissertation, Technische Universität Darmstadt, 2006.

[BH91]   J.R. Brauer and F. Hirtenfelder. Anisotropic materials in electromagnetic finite element analysis. *International Conference on Computation in Electromagnetics, 1991*, pages 144–147, 1991.

[BIG03]   A. Ben-Israel and T.N.E. Greville. *Generalized inverses: theory and applications*. CMS books in mathematics. Springer, 2003.

[BMR00]   P.I. Barton, W.S. Martinson, and G. Reissig. Differential-Algebraic Equations of Index 1 May Have an Arbitrarily High Structural Index. *SIAM journal on Scientific Computing*, 21(6):1987–1990, 2000.

[BO71]   T.L. Boullion and P.L. Odell. *Generalized Inverse Matrices.* Interscience, 1971.

[Bod07]   M. Bodestedt. *Perturbation Analysis of Refined Models in Circuit Simulation.* Dissertation, Technische Universität Berlin, 2007.

[Bos88]   A. Bossavit. Whitney forms: a class of finite elements for three-dimensional computations in electromagnetism. *IEE Proceedings*, 135(8):493–500, 1988.

[Bos98]   A. Bossavit. *Computational electromagnetism: variational formulations, complementarity, edge elements*, volume 2. Academic Press, Boston, 1998.

[Bos01]   A. Bossavit. Stiff problems in eddy-current theory and the regularization of Maxwell's equations. *IEEE Transactions on Magnetics*, 37(5):3542–3545, 2001.

[BP89]   O. Bíró and K. Preis. On the use of the magnetic vector potential in the finite-element analysis of three-dimensional eddy currents. *IEEE Transactions on Magnetics*, 25(4):3145–3159, 1989.

[BST10]   S. Baumanns, M. Selva Soto, and C. Tischendorf. Consistent initialization for coupled circuit-device simulation. In J. Roos and L.R.J. Costa, editors, *Scientific Computing in Electrical Engineering SCEE 2008*, number 14 in Mathematics in Industry, pages 297–304, Berlin, 2010. Springer.

[BST12]   S. Baumanns, M. Selva Soto, and C. Tischendorf. A monolithic PDAE approach for coupling circuit and semiconductor device simulation. *In preparation*, 2012.

[BT10]   S. Baumanns and C. Tischendorf. Simulation Aspects of Circuit DAEs Including Memristors. In G. Psihoyios T.E. Simos and C. Tsitouras, editors, *American Institute of Physics Conference series*, volume 1281 of *American Institute of Physics Conference series*, pages 996–999, 2010.

[But03]   J.C. Butcher. Miniature 21: R gave me a DAE underneath the Linden tree. Technical report, University of Auckland, Auckland, 2003.

[Cam87]   S.L. Campbell. A general form for solvable linear time varying singular systems of differential equations. *SIAM journal on Mathematical Analysis*, 18(4):1101–1115, 1987.

[CDK87]   L.O. Chua, C.A. Desoer, and E.S. Kuh. *Linear and nonlinear circuits.* McGraw-Hill, Singapore, 1987.

[Chu71]   L.O. Chua. Memristor-The missing circuit element. *IEEE Transactions on Circuit Theory*, 18(5):507–519, 1971.

[Chu11]   L.O. Chua. Resistance switching memories are memristors. *Applied Physics A: Materials Science & Processing*, 102(4):765–783, 2011.

[CK97]   Q. Chen and A. Konrad. A review of finite element open boundary techniques for static and quasi-static electromagnetic field problems. *IEEE Transactions on Magnetics*, 33(1):663–676, 1997.

[CL75]   L.O. Chua and P.M. Lin. *Computer-aided analysis of electronic circuits: algorithms and computational techniques*. Prentice-Hall series in electrical and computer engineering. Prentice-Hall, 1975.

[Cle98]   M. Clemens. *Zur numerischen Berechnung zeitlich langsam veränderlicher elektromagnetischer Felder mit der Finiten-Integrations-Methode*. Dissertation, Technische Universität Darmstadt, 1998.

[Cle05]   M. Clemens. Large systems of equations in a discrete electromagnetism: formulations and numerical algorithms. *IEE Proceedings - Science, Measurement and Technology*, 152(2):50–72, 2005.

[CMSW11]   Q. Chen, P. Meuris, W. Schoenmaker, and N. Wong. An Effective Formulation of Coupled Electromagnetic-Semiconductor Simulation for Extremely High Frequency Onwards. *IEEE Transactions on Computer-Aided Design of Integrated Circuits and Systems*, 30(6):866–876, 2011.

[CPD09]   L.O. Chua, Y.V. Pershin, and M. Di Ventra. Circuit elements with memory: memristors, memcapacitors, and meminductors. *Proceedings of the IEEE*, 97(10):1717–1724, 2009.

[CSvW96]   M. Clemens, R. Schuhmann, U. van Rienen, and T. Weiland. Modern krylov subspace methods in electromagnetic field computation using the finite integration theory. *Applied Computational Electromagnetics Society*, 11(1):70–84, 1996.

[Cul09]   M. Culpo. *Numerical algorithms for system level electro-thermal simulation*. Dissertation, Bergischen Universität Wuppertal, 2009.

[CW99]   M. Clemens and T. Weiland. Transient eddy-current calculation with the FI-method. *IEEE Transactions on Magnetics*, 35(3):1163–1166, 1999.

[CW01a]   M. Clemens and T. Weiland. Discrete electromagnetics: Maxwell's equation tailored to numerical simulations. *International Compumag Society Newsletter*, 8:13–20, 2001.

[CW01b]   M. Clemens and T. Weiland. Discrete Electromagnetism with the Finite Integration Technique. *Progress In Electromagnetics Research*, 32:65–87, 2001.

[CW02]   M. Clemens and T. Weiland. Regularization of eddy-current formulations using discrete grad-div operators. *IEEE Transactions on Magnetics*, 38(2):569–572, 2002.

[Dau10]   E. Dautbegovic. private communication, 2010.

[DF06]   G. Denk and U. Feldmann. Circuit Simulation for Nanoelectronics. In A.M. Anile, G. Alì, and G. Mascali, editors, *Scientific Computing in Electrical Engineering SCEE 2004*, number 9 in Mathematics in Industry, pages 11–26, Berlin, 2006. Springer.

[DHW04]   H. De Gersem, K. Hameyer, and T. Weiland. Field-Circuit Coupled Models in Electromagnetic Simulation. *Journal of Computational and Applied Mathematics*, 168(1-2):125–133, 2004.

[Die05]   R. Diestel. *Graph Theory*. Graduate Texts in Mathematics. Springer, 2005.

[DK84]   C.A. Desoer and E.S. Kuh. *Basic Circuit Theory*. International student edition. McGraw-Hill, 1984.

[DMW08]   H. De Gersem, I. Munteanu, and T. Weiland. Construction of Differential Material Matrices for the Orthogonal Finite-Integration Technique With Nonlinear Materials. *IEEE Transactions on Magnetics*, 44(6):710–713, 2008.

[Doh92]   M. Dohlus. *Ein Beitrag zur numerischen Berechnung elektromagnetischer Felder im Zeitbereich*. Dissertation, Technische Universität Darmstadt, 1992.

[DW04]   H. De Gersem and T. Weiland. Field-Circuit Coupling for Time-Harmonic Models Discretized by the Finite Integration Technique. *IEEE Transactions on Magnetics*, 40(2):1334–1337, 2004.

[EFM+03]   D. Estévez Schwarz, U. Feldmann, R. März, S. Sturtzel, and C. Tischendorf. Finding Beneficial DAE Structures in Circuit Simulation. In W. Jäger and H.-J. Krebs, editors, *Mathematics-Key Technology for the Future: Joint Projects Between Universities and Industry*, pages 413–428. Springer, Berlin, 2003.

[ESF98]   E. Eich-Soellner and C. Führer. *Numerical Methods in Multibody Systems*. Lecture Notes in Mathematics. Teubner, Stuttgart, 1998.

[Est00]   D. Estévez Schwarz. *Consistent initialization for index-2 differential algebraic equations and its application to circuit simulation*. Dissertation, Humboldt-Universität zu Berlin, 2000.

[ET00]   D. Estévez Schwarz and C. Tischendorf. Structural analysis of electric circuits and consequences for MNA. *International journal of Circuit Theory and Applications*, 28(2):131–162, 2000.

[Eva10]   L.C. Evans. *Partial Differential Equations*. Graduate Studies in Mathematics. American Mathematical Society, 2010.

[FG99]   U. Feldmann and M. Günther. CAD-based electric-circuit modeling in industry I: mathematical structure and index of network equations. *Surveys on Mathematics for Industry*, 8(2):97–129, 1999.

[FY10]   W. Fei and H. Yu. A new modified nodal analysis for nano-scale memristor circuit simulation. In *Proceedings of IEEE International Symposium on Circuits and Systems*, pages 3148–3151, 2010.

[Gan71]   F.R. Gantmacher. *Matrizenrechnung*, volume 1 and 2 of *Hochschulbücher für Mathematik*. Deutscher Verlag der Wissenschaft, Berlin, 1971.

[Gea06]   C.W. Gear. Towards explicit methods for differential algebraic equations. *BIT Numerical Mathematics*, 46(3):505–514, 2006.

[GGL85]   C.W. Gear, G.K. Gupta, and B. Leimkuhler. Automatic integration of Euler-Lagrange equations with constraints. *Journal of Computational and Applied Mathematics*, 12-13(0):77–90, 1985.

[GHM92]   E. Griepentrog, M. Hanke, and R. März. Toward a better understanding of differential algebraic equations (Introductory survey). Number 2 in Berliner Seminar on Differential-Algebraic Equations. Seminar Notes. Humboldt-Universität zu Berlin, Institut für Mathematik, 1992.

[GM86]   E. Griepentrog and R. März. *Differential-Algebraic Equations and Their Numerical Treatment*. Teubner, Leipzig, 1986.

[Göd10]   N. Gödel. *Numerische Simulation hochfrequenter elektromagnetischer Felder durch die Discontinuous Galerkin Finite Elemente Methode*. Dissertation, Helmut-Schmidt-Universität/Universität der Bundeswehr Hamburg, 2010.

[GP83]   C.W. Gear and L.R. Petzold. Differential/algebraic systems and matrix pencils. In B. Kågström and A. Ruhe, editors, *Matrix Pencils*, volume 973 of *Lecture Notes in Mathematics*, pages 75–89. Springer, Berlin/Heidelberg, 1983.

[GP84]   C.W. Gear and L.R. Petzold. ODE Methods for the Solution of Differential/Algebraic Systems. *SIAM journal on Numerical Analysis*, 21(4):716–728, 1984.

[Gün01]   M. Günther. *Partielle differential-algebraische Systeme in der numerischen Zeitbereichsanalyse elektrischer Schaltungen*. Number 343 in Fortschritt-Berichte VDI, Reihe 20. VDI Verlag, Düsseldorf, 2001. Habilitation.

[Hip98]   R. Hiptmair. Multigrid method for Maxwell's equations. *SIAM Journal on Numerical Analysis*, 36(1):204–225, 1998.

[HLR89]   E. Hairer, C. Lubich, and M. Roche. *The Numerical Solution of Differential-Algebraic Systems by Runge-Kutta Methods.* Springer, New York, 1989.

[HM76]   A.Y. Hannalla and D.C. MacDonald. Numerical analysis of transient field problems in electrical machines. *Proceedings of the Institution of Electrical Engineers*, 123(9):893–898, 1976.

[HM89]   H.A. Haus and J.R. Melcher. *Electromagnetic Fields and Energy.* Prentice Hall, 1989.

[HM04]   I. Higueras and R. März. Differential algebraic equations with properly stated leading terms. *Computers & Mathematics with Applications*, 48(1-2):215–235, 2004.

[HMT03a]   I. Higueras, R. März, and C. Tischendorf. Stability Preserving Integeration of Index-1 DAEs. *Applied Numerical Mathematics*, 45(2-3):175–200, 2003.

[HMT03b]   I. Higueras, R. März, and C. Tischendorf. Stability Preserving Integeration of Index-2 DAEs. *Applied Numerical Mathematics*, 45(2-3):201–229, 2003.

[HNW02]   E. Hairer, S.P. Nørsett, and G. Wanner. *Solving Ordinary Differential Equations II: Stiff and Differential-Algebraic Problems.* Springer series in Computational Mathematics. Springer, Berlin, 2 edition, 2002.

[HW05]   H. Hoeber and A. Wachter. *Compendium of Theoretical Physics.* 2005.

[Jac98]   J.D. Jackson. *Classical Electrodynamics.* John Wiley & Sons, New York, 3rd edition, 1998.

[Jan12a]   L. Jansen. *Higher Index DAEs: Unique Solvability and Convergence of Implicit and Half-Explicit Multistep Methods.* In preparation, 2012.

[Jan12b]   L. Jansen. private communication, 2012.

[JMT12]   L. Jansen, M. Matthes, and C. Tischendorf. Global Unique Solvability of Nonlinear Index-1 DAEs with Monotonicity Properties. *In preparation*, 2012.

[KKS10]   S.-M. Kang, K. Kim, and S. Shin. Compact Models for Memristors Based on Charge-Flux Constitutive Relationships. *IEEE Transactions on Computer-Aided Design of Integrated Circuits and Systems*, 29(4):590–598, 2010.

[KM94]   P. Kunkel and V. Mehrmann. Canonical forms for linear differential-algebraic equations with variable coefficients. *journal of computational and applied mathematics*, 56(3):225–251, 1994.

[KM06]   P. Kunkel and V. Mehrmann. *Differential-algebraic equations: analysis and numerical solution.* EMS textbooks in mathematics. European Mathematical Society, 2006.

185

[KMST93] A. Konrad, G. Meunier, J.C. Sabonnadière, and I.A. Tsukerman. Coupled field-circuit problems: trends and accomplishments. *IEEE Transactions on Magnetics*, 29(2):1701–1704, 1993.

[Koc09] Stefan Koch. *Quasistatische Feldsimulationen auf der Basis von Finiten Elementen und Spektralmethoden in der Anwendung auf supraleitende Magnet.* Dissertation, Technische Universität Darmstadt, 2009.

[Krü00] H. Krüger. *Zur numerischen Berechnung transienter elektromagnetischer Felder in gyrotropen Materialien.* Dissertation, Technische Universität Darmstadt, 2000.

[LMT13] R. Lamour, R. März, and C. Tischendorf. *Differential-Algebraic Equations: A Projector Based Analysis.* to appear in Springer, 2013.

[Mär96] R. März. Managing the drift-off in numerical index-2 differential algebraic equations by projected defect corrections. Number 32 in Preprints aus dem Institut für Mathematik. Humboldt-Universität zu Berlin, Institut für Mathematik, 1996.

[Mär01] R. März. Nonlinear differential-algebraic equations with properly formulated leading term. Technical Report 01-3, Humboldt-Universität zu Berlin, 2001.

[Mär02] R. März. The index of linear differential algebraic equations with properly stated leading terms. *Results in Mathematics*, 42(3-4):308–338, 2002.

[Mär03] R. März. Differential Algebraic Systems with Properly Stated Leading Term and MNA Equations. In K. Anstreich, R. Bulirsch, A. Gilg, and P. Rentrop, editors, *Modelling, Simulation and Optimization of Integrated Circuits*, pages 135–151, Berlin, 2003. Birkhäuser.

[Mat12] M. Matthes. *Numerical Analysis of Nonlinear Partial Differential Algebraic Equations: A Coupled and an Abstract Systems Approach.* In preparation, 2012.

[Max64] J.C. Maxwell. A Dynamical Theory of the Electromagnetic Field. *Royal Society Transactions*, 155:459 – 512, 1864.

[Meh12] V. Mehrmann. Index concepts for differential-algebraic equations. Number 2 in Preprints aus dem Institut für Mathematik. Technische Universität zu Berlin, Institut für Mathematik, 2012.

[Men11] M. Menrath. *Stability criteria for nonlinear fully implicit differential-algebraic systems.* Dissertation, Universität zu Köln, 2011.

[MMS01]   W. Magnus, P. Meuris, and W. Schoenmaker. Strategy for electromagnetic interconnect modeling. *IEEE Transactions on Computer-Aided Design of Integrated Circuits and Systems*, 20(6):753–762, 2001.

[Mon03]   P. Monk. *Finite element methods for Maxwell's equations*. Numerical mathematics and scientific computation. Clarendon Press, 2003.

[MR99]   R. März and A.R. Rodríguez Santiesteban. Analyzing the stability behaviour of DAE solutions and their approximations. Number 2 in Preprints aus dem Institut für Mathematik. Humboldt-Universität zu Berlin, Institut für Mathematik, 1999.

[MRT05]   R.M.M. Mattheij, S.W. Rienstra, and J.H.M. Thije Boonkamp. *Partial differential equations - Modeling, analysis, computation*. Philadelphia, 2005.

[MW07]   I. Munteanu and T. Weiland. RF & Microwave Simulation with the Finite Integration Technique - From Component to System Design. In G. Ciuprina and D. Ioan, editors, *Scientific Computing in Electrical Engineering SCEE 2006*, number 11 in Mathematics in Industry, pages 247–260. Springer, 2007.

[Ned80]   J.C. Nedelec. Mixed finite elements in $\mathbb{R}^3$. *Numerische Mathematik*, 35:315–341, 1980.

[OST01]   C.W. Oosterlee, A. Schüller, and U. Trottenberg. *Multigrid*. Academic Press, 2001.

[Pet82]   L.R. Petzold. Differential/algebraic equations are not ODEs. *SIAM journal on Scientific Computing*, 3(3):367–384, 1982.

[Pry01]   J. Pryce. A Simple Structural Analysis Method for DAEs. *BIT Numerical Mathematics*, 41(2):364–394, 2001.

[Rei06]   T. Reis. *Systems Theoretic Aspects of PDAEs and Applications to Electrical Circuits*. Dissertation, Technische Universität Kaiserslautern, 2006.

[Ria08]   R. Riaza. *Differential-Algebraic Systems: Analytical Aspects and Circuit Applications*. World Scientifc, 2008.

[Ria10]   R. Riaza. First order devices, hybrid memristors, and the frontiers of nonlinear circuit theory. *arXiv*, 2010.

[Ria11]   R. Riaza. Dynamical properties of electrical circuits with fully nonlinear memristors. *Nonlinear Analysis: Real World Applications*, 12(6):3674–3686, 2011.

[Rod00]   A.R. Rodríguez Santiesteban. *Asymptotische Stabilität von Index-2 Algebro-Differentialgleichungen und ihren Diskretisierungen*. Dissertation, Humboldt Universität zu Berlin, 2000.

[RR90]   P.J. Rabier and W.C. Rheinboldt. *A General Existence and Uniqueness Theorem for Implicit Differential-Algebraic Equations.* Defense Technical Information Center, 1990.

[RR02]   P.J. Rabier and W.C. Rheinboldt. *Techniques of scientific computing (part 4) - theoretical and numerical analysis of differential-algebraic equations.*, volume 8 of *Handbook of numerical analysis.* Elsevier, Amsterdam, 2002.

[RT11]   R. Riaza and C. Tischendorf. Semistate models of electrical circuits including memristors. *International journal of Circuit Theory and Applications*, 39(6):607–627, 2011.

[Saa03]   Y. Saad. *Iterative methods for sparse linear systems.* SIAM, 2003.

[Sch03]   S. Schulz. Four lectures on differential algebraic equations. Technical Report 497, University of Auckland, New Zealand, 2003.

[Sch11]   S. Schöps. *Multiscale Modeling and Multirate Time-Integration of Field-/Circuit Coupled Problems.* Dissertation, Bergischen Universität Wuppertal, 2011.

[Sel06]   M. Selva Soto. An Index Analysis from Coupled Circuit and Device Simulation. In A.M. Anile, G. Alì, and G. Mascali, editors, *Scientific Computing in Electrical Engineering SCEE 2004*, number 9 in Mathematics in Industry, pages 121–128, Berlin, 2006. Springer.

[SSSW08]   G.S. Snider, D.R. Stewart, D.B. Strukov, and R.S. Williams. The missing memristor found. *Nature*, 453:80–83, 2008.

[ST05]   M. Selva Soto and C. Tischendorf. Numerical Analysis of DAEs from coupled circuit and semiconductor simulation. *Applied Numerical Mathematics*, 53(2-4):471–488, 2005.

[StM05]   W.H.A. Schilders and E.J.W. ter Maten, editors. *Special volume: numerical methods in electromagnetics*, volume 13 of *Handbook of numerical analysis.* Elsevier, Amsterdam, 2005.

[Str07]   J.A. Stratton. *Electromagnetic Theory.* IEEE Press series on Electromagnetic Wave Theory. John Wiley & Sons, 2007.

[SV93]   K. Singhal and J. Vlach. *Computer methods for cirucit analysis and design.* Springer, 1993.

[SW96]   S.R. Sanders and D.M. Wolf. Multiparameter homotopy methods for finding DC operating points of nonlinear circuits. *IEEE Transactions on Circuits and Systems I: Fundamental Theory and Applications*, 43(10):824–838, 1996.

[Ter43]   F.E. Terman. *Radio engineer's handbook.* McGraw-Hill, 1943.

[Tis96]   C. Tischendorf. *Solution of index-2 differential algebraic equations and its application in circuit simulation.* Dissertation, Humboldt-Universität zu Berlin, 1996.

[Tis99]   C. Tischendorf. Topological index calculation of DAEs in circuit simulation. *Surveys on Mathematics for Industry,* 8(3-4):187–199, 1999.

[Tis01]   C. Tischendorf. Model Design Criteria for Integrated Circuits to Have a Unique Solution and Good Numerical Properties. In U. van Rienen, M. Günther, and D. Hecht, editors, *Scientific Computing in Electrical Engineering SCEE 2000,* Mathematics in Industry, pages 179–198, Berlin, 2001. Springer.

[Tis04]   C. Tischendorf. *Coupled Systems of Differential Algebraic and Partial Differential Equations in Circuit and Device Simulation. Modeling and Numerical Analysis.* Habilitation, Humboldt-Universität zu Berlin, 2004.

[Ton95]   E. Tonti. On the Geometrical Structure of Electromagnetism. In G. Ferrarese, editor, *Gravitation, Electromagnetism and Geometrical Structures,* pages 281–308, Bologna, 1995. Pitagora Editrice.

[Ton01]   E. Tonti. A direct discrete formulation of field laws: The cell method. *Computer Modeling in Engineering and Sciences,* 2(2):237–258, 2001.

[Tsu02]   I.A. Tsukerman. Finite Element Differential-Algebraic Systems for Eddy Current Problems. *Numerical Algorithms,* 31(1):319–335, 2002.

[TW96]   P. Thoma and T. Weiland. A consistent subgridding scheme for the finite difference time domain method. *International journal of Numerical Modelling: Electronic Networks, Devices and Fields,* 9(5):359–374, 1996.

[Vla94]   A. Vladimirescu. *The spice book.* John Wiley & Sons, 1994.

[Voi06]   S. Voigtmann. *General linear methods for integrated circuit design.* Dissertation, Humboldt-Universität zu Berlin, 2006.

[Wei77]   T. Weiland. A discretization model for the solution of Maxwell's equations for six-component fields. *Archiv Elektronik und Übertragungstechnik,* 31(3):116–120, 1977.

[Wei96]   T. Weiland. Time Domain Electromagnetic Field Computation with Finite Difference Methods. *International Journal of Numerical Modelling: Electronic Networks, Devices and Fields,* 9(4):295 – 319, 1996.

[Yee66]   K.S. Yee. Numerical Solution of Initial Boundary Value Problems Involving Maxwell's Equations in Isotropic Media. *IEEE Transactions on Antennas and Propagation,* 14(3):302–307, 1966.

189

Index